Encyclopaedia of
Animal Cell and Tissue Culture

Encyclopaedia of Animal Cell and Tissue Culture

Volume 1

Editor:
Dr. Anil M. Mane

AGROTECH PRESS
JAIPUR ▪ NEW DELHI

AGROTECH PRESS

Regd. Off. "Bony Residency", Gate No. 2
Opp. Tilak Public School, Vishweswariya Nagar,
Gopalpura Road, Jaipur - 302018, India
Ph: 0141 - 2761280, Telefax: 2761381
E-mail : agrotechpress@gmail.com

Delhi Office: 102, 1st Floor, Satyam House,
4327/3, Ansari Road, Darya Ganj,
New Delhi - 110002, India
Ph: 011-45652440
visit us: www.agrotechpress.com

Distributed By
KEYA INTERNATIONAL PRESS LLP
E-mail: keyainternationalpressllp@gmail.com

Published in India, 2025

ISBN : 9789384568276 (3 Vols. Set)

Typeset by Thomson Press

Printed and bound in India by Thomson Press Ltd.,
New Delhi

Contents

Preface

Cell culture is the complex process by which cells are grown under controlled conditions, generally outside of their natural environment. In practice, the term "cell culture" now refers to the culturing of cells derived from multi-cellular eukaryotes, especially animal cells. However, there are also cultures of plants, fungi and microbes, including viruses, bacteria and protists. The historical development and methods of cell culture are closely interrelated to those of tissue culture and organ culture.

Animal cell culture became a common laboratory technique in the mid-1900s, but the concept of maintaining live cell lines separated from their original tissue source was discovered in the 19th century. Cells can be isolated from tissues for *ex vivo* culture in several ways. Cells can be easily purified from blood; however, only the white cells are capable of growth in culture. Mononuclear cells can be released from soft tissues by enzymatic digestion with enzymes such as collagenase, trypsin, or pronase, which break down the extracellular matrix. Alternatively, pieces of tissue can be placed in growth media, and the cells that grow out are available for culture. This method is known as explant culture. Cells that are cultured directly from a subject are known as primary cells. With the exception of some derived from tumours, most primary cell cultures have limited lifespan. After a certain number of population doublings, cells undergo the process of senescence and stop dividing, while generally retaining viability. An established or immortalized cell line has acquired the ability to proliferate indefinitely either through random mutation or deliberate modification, such as artificial expression of the telomerase gene. Numerous cell lines are well established as representative of particular cell types. Among the essential requirements for animal cell culture are special incubators to maintain the levels of oxygen, carbon dioxide, temperature, humidity as present in the animal's body. The synthetic media with vitamins, amino acids and fetal calf serum. Following parameters are essential for successful animal cell culture: In most of the mammalian cell cultures, the temperature is maintained at 370C in the incubators as the body temperature of Homo sapiens is 370C. The culture media is prepared in such a way that it provides- The optimum conditions of factors like pH, osmotic pressure, etc. It should contain chemical constituents which the cells or tissues are incapable of synthesizing. Generally the media is the mixture of inorganic salts and other nutrients capable of sustaining cells in culture such as amino acids, fatty acids, sugars, ions, trace elements, vitamins, cofactors, and ions. Glucose is added as energy source-it's concentration varying depending on the requirement. Phenol Red is added as a pH indicator of the medium. There are two types of media used for culture of animal cells and tissues- the natural media and the synthesized media. The natural media are the natural sources of nutrient sufficient for growth and proliferation of animal cells and tissues. Syntheic media are prepared artificially by adding several organic and inorganic nutrients, vitamins, salts, serum proteins, carbohydrates, cofactors etc. Different types of synthetic media can be prepared for a variety of cells and tissues to be cultured. Synthetic media are of two types- Serum containing media (media containing serum) and serum- free media (media with out serum). Examples of some media are: minimal essential medium (MEM), RPMI 1640 medium, CMRL 1066, F12 etc. Most media maintain the pH between 7 and 7.4. A pH below 6.8 inhibits cell growth. The optimum pH is essential to maintain the proper ion balance, optimal functioning of cellular enzymes and binding of hormones and growth factors to cell surface receptors in the cell cultures. The regulation of pH is done using a variety of buffering systems. Most media use a bicarbonate-CO2 system as its major component. A change in osmolality can affect cell growth and function. Salt, Glucose and Amino acids in the growth media determine the osmolality of the medium. All commercial media are formulated in such a way that their final osmolality is around 300 mOsm.

This book will be extremely useful to all persons who are directly or indirectly involved in cell culture work for various biological experiments. Finally, students and examinees can enrich their knowledge on cell culture from the book and can face any challenge easily and confidently.

—Editor

1

Human Stem Cell

Amniotic fluid surrounding the developing fetus contains cells shed by the fetus and is regularly collected from pregnant women during amniocentesis. In 2003, researchers identified a subset of cells in amniotic fluid that express Oct-4, a marker for pluripotent human stem cells that is expressed in ES cells and embryonic germ cells. Since then, investigators have shown that amniotic fluid stem cells can differentiate into cells of all three embryonic germ layers and that these cells do not form tumours in vivo.

For example, Anthony Atala and colleagues at the Wake Forest University have recently generated non-embryonic stem cell lines from cells found in human and rat amniotic fluid. They named these cells amniotic fluid-derived stem cells (AFS). Experiments demonstrate that AFS can produce cells that originate from each of the three embryonic germ layers, and the self-renewing cells maintained the normal number of chromosomes after a prolonged period in culture. However, undifferentiated AFS did not produce all of the proteins expected of pluripotent cells, and they were not capable of forming a teratoma. The scientists developed *in vitro* conditions that enabled AFS to produce nerve cells, liver cells, and bone-forming cells.

AFS-derived human nerve cells could make proteins typical of specialised nerve cells and were able to integrate into a mouse brain and survive for at least two months. Cultured AFS-derived human liver cells secreted urea and made proteins characteristic of normal human liver cells. Cultured AFS-derived human bone cells made proteins expected of human bone cells and formed bone in mice when seeded onto scaffolds and implanted under the mouse's skin. Although scientists do not yet know how many different cell types AFS can generate, AFS may one day allow researchers to establish a bank of cells for transplantation into humans.

Strategies to "Reprogram" Non-Pluripotent Cells Become Pluripotent Cells

An alternative to searching for an existing population of stem cells is to create a new one from a population of non-pluripotent cells. This strategy, which may or may not involve the creation of an embryo, is known as "reprogramming." This section will summarize reprogramming approaches, including several recent breakthroughs in the field..

Reprogramming Through Somatic Cell Nuclear Transfer (SCNT)

In SCNT, human oocytes (eggs) are collected from a volunteer donor who has taken drugs that stimulate the production of more than one oocyte during the menstrual cycle. Scientists then remove the nucleus from the donated oocyte and replace it with the nucleus from a somatic cell, a differentiated adult cell from elsewhere in the body. The oocyte with the newly-transferred nucleus is then stimulated to develop. The oocyte may develop only if the transplanted nucleus is returned to the pluripotent state by factors present in the oocyte cytoplasm. This alteration in the state of the mature nucleus is called nuclear reprogramming. When development progresses to the blastocyst stage, the ICM is removed and placed into culture in an attempt to establish a pluripotent stem cell line. To date, the technique has been successfully demonstrated in two primates: macaque monkeys and humans.

However, successful SCNT creates an embryo-like entity, thereby raising the ethical issues that confront the use of spare IVF embryos. However, pluripotent cell lines created by embryos generated by SCNT offer several advantages over ES cells. First, the nuclear genes of such a pluripotent cell line will be identical to the genes in the donor nucleus. If the nucleus comes from a cell that carries a mutation underlying a human genetic disease such as Huntington's disease, then all cells derived from the pluripotent cell line will carry this mutation. In this case, the SCNT procedure would enable the development of cellular models of human genetic disease that can inform our understanding of the biology of disease and facilitate development of drugs to slow or halt disease progression. Alternatively, if the cell providing the donor nucleus comes from a specific patient, all cells derived from the resulting pluripotent cell line will be genetically matched to the patient with respect to the nuclear genome. If these cells were used in transplantation therapy, the likelihood that the patient's immune system would recognize the transplanted cells as foreign and initiate tissue rejection would be reduced. However, because mitochondria also contain DNA, the donor oocyte will be the source of the mitochondrial genome, which is likely to carry mitochondrial gene differences from the patient which may still lead to tissue rejection. A technique reported in 2007 by Dr. Kevin Eggan and colleagues at Harvard University may expand scientists' options when trying to "reprogram" an adult cell's DNA. Previously, successful SCNT relied upon the use of an unfertilized egg. Now, the Harvard scientists have demonstrated that by using a drug to stop cell division in a fertilized mouse egg (zygote) during mitosis, they can successfully reprogram an adult mouse skin cell by taking advantage of the "reprogramming factors" that are active in the zygote at mitosis. They removed the chromosomes from the single-celled zygote's nucleus and replaced them with the adult donor cell's chromosomes. The active reprogramming factors present in the zygote turned genes on and off in the adult donor chromosomes, to make them behave like the chromosomes of a normally fertilized zygote. After the zygote was stimulated to divide, the cloned mouse embryo developed to the blastocyst stage, and the scientists were able to harvest embryonic stem cells from the resulting blastocyst. When the scientists applied their new method to abnormal mouse zygotes, they succeeded at reprogramming adult mouse skin cells and harvesting stem cells. If this technique can be repeated with abnormal human zygotes created in excess after IVF procedures, scientists could use them for research instead of discarding them as medical waste.

Reprogramming through Altered Nuclear Transfer (ANT)

Altered nuclear transfer is a variation on standard SCNT that proposes to create patient-specific stem cells without destroying an embryo. In ANT, scientists turn off a gene needed for implantation in the uterus (Cdx2) in the patient cell nucleus before it is transferred into the donor egg. In 2006, Dr. Rudolph Jaenisch and colleagues at MIT demonstrated that ANT can be carried out in mice. Mouse ANT entities whose Cdx2 gene is switched off are unable to implant in the uterus and do not survive to birth. Although ANT has been used to create viable stem cell lines capable of producing almost all cell types, the authors point out that this technique must still be tested with monkey and human embryos. Moreover, the manipulation needed to control Cdx2 expression introduces another logistical hurdle that may complicate the use of ANT to derive embryonic stem cells. Proponents of ANT, such as William Hurlbut of the Stanford University Medical Centre, suggest that the entity created by ANT is not a true embryo because it cannot implant in the uterus. However, the technique is highly controversial, and its ethical implications remain a source of current debate.

Reprogramming through Cell Fusion

In 2005, Kevin Eggan and colleagues at Harvard University reported that they had fused cultured adult human skin cells with hESCs. The resulting "hybrid" cells featured many characteristics of hESCs, including a similar manner of growth and division and the manufacture of proteins typically produced by hESCs. Some factor(s) within the hESCs enabled them to "reprogram" the adult skin cells to behave as hESCs. However, these cells raised a significant technical barrier to clinical use. Because fused cells are tetraploid (they contain four copies of the cellular DNA rather than the normal two copies), scientists would need to develop a method to remove the extra DNA without eliminating their hESC-like properties. The fusion method serves as a useful model system for studying how stem cells "reprogram" adult cells to have properties of pluripotent cells. However, if the reprogramming technique could be carried out without the fusion strategy, a powerful avenue for creating patient-specific stem cells without using human eggs could be developed.

Induced Pluripotent Stem Cells (iPSCs): Reprogramming Adult Somatic Cells to Become Pluripotent Stem Cells

In 2007, two independent research groups published manuscripts that described successful genetic reprogramming of human adult somatic cells into pluripotent human stem cells. Although some technical limitations remain, this strategy suggests a promising new avenue for generating pluripotent cell lines that can inform drug development, models of disease, and ultimately, transplantation medicine. These experiments, which are discussed below, were breakthroughs because they used adult somatic cells to create pluripotent stem cells that featured hallmarks of ES cells.

In 2006, Shinya Yamanaka and colleagues at Kyoto University reported that they could use a retroviral expression vector to introduce four important stem cell factors into adult mouse cells and reprogram them to behave like ES cells. They called the reprogrammed cells "iPSCs," for induced pluripotent stem cells. However, iPSCs produced using the original technique failed to produce sperm and egg cells when injected into an early mouse blastocyst and did not make certain critical DNA changes. These researchers then modified the technique to select for iPSCs that can produce sperm and eggs, results that have since been reproduced by Rudolph Jaenisch and colleagues at the Massachusetts Institute of Technology (MIT). In addition, the MIT scientists determined that iPSCs DNA is modified in a manner similar to ES cells, and important stem cell genes are expressed at similar levels. They also demonstrated that iPSCs injected into an early mouse blastocyst can produce all cell types within the developing embryo, and such embryos can complete gestation and are born alive.

Once these research advances were made in mice, they suggested that similar techniques might be used to reprogram adult human cells. In 2007, Yamanaka and coworkers reported that introducing the same four genetic factors that reprogrammed the mouse cells into adult human dermal fibroblasts reprogrammed the cells into human iPSCs. These iPSCs were similar to human ES cells in numerous ways, including morphology, proliferative capacity, expression of cell surface antigens, and gene expression. Moreover, the cells could differentiate into cell types from the three embryonic germ layers both in vitro and in teratoma assays. Concurrent with the Yamanaka report, James Thomson and coworkers at the University of Wisconsin published a separate manuscript that detailed the creation of human iPSCs through somatic cell reprogramming using four genetic factors (two of which were in common with the Yamanaka report). The cells generated by the Thomson group met all defining criteria for ES cells, with the exception that they were not derived from embryos.

These breakthroughs have spurred interest in the field of iPSCs research. In early 2008, investigators at the Massachusetts General Hospital and the University of California, Los Angeles reported generating reprogrammed cells. As scientists explore the mechanisms that govern reprogramming, it is anticipated that more reports will be forthcoming in this emerging area. Although these reprogramming methods require the use of a virus, non-viral strategies may also be possible in the future. In any case, these approaches have created powerful new tools to enable the "de-differentation" of cells that scientists had previously believed to be terminally differentiated. Although further study is warranted to determine if iPS and ES cells differ in clinically significant ways, these breakthrough reports suggest that reprogramming is a promising strategy for future clinical applications. Induced pluripotent cells offer the obvious advantage that they are not derived from embryonic tissues, thereby circumventing the ethical issues that surround use of these materials. Successful reprogramming of adult somatic cells could also lead to the development of stem cell lines from patients who suffer from genetically-based diseases, such as Huntington's Disease, spinal muscular atrophy, muscular dystrophy, and thalessemia. These lines would be invaluable research tools to understand the mechanisms of these diseases and to test potential drug treatments. Additionally, reprogrammed cells could potentially be used to repair damaged tissues; patient-specific cell lines could greatly reduce the concerns of immune rejection that are prevalent with many transplantation strategies. However, several technical hurdles must be overcome before iPSCs can be used in humans. For example, in preliminary experiments with mice, the virus used to introduce the stem cell factors sometimes caused cancers. The viral vectors used in these experiments will have to be selected carefully and tested fully to verify that they do not integrate into the genome, thereby harbouring the potential to introduce genetic mutations at their site of insertion. This represents a significant concern that must be addressed before the technique can lead to useful treatments for humans. However, this strategy identifies a method for creating pluripotent stem cells that, together with studies of other types of pluripotent stem cells, will help researchers learn how to reprogramme cells to repair damaged tissues in the human body.

Other Sources of Pluripotent and/or Multipotent Cells

Stem cell research is a rapidly evolving field, and researchers continue to isolate new pluripotent cells and create additional cell lines. This section briefly reviews other sources of pluripotent cells and the implications that their discovery may have on future research.

Epiblast Cells

While rodent and human ES cells are pluripotent, they maintain their respective pluripotencies through different molecular signalling pathways. It is not known why these differences ex7ist. Recently, several research groups have reported the generation of stable, pluripotent cell lines from mouse and rat epiblast, a tissue of the post-implantation embryo that ultimately generates the embryo proper. These cells are distinct from mouse ES cells in terms of the signals that control their differentiation. However, the cells share patterns of gene expression and signalling responses with human ES cells. The establishment of epiblast cell lines can therefore provide insight into the distinctions between pluripotent cells from different species and illuminate ways that pluripotent cells pursue distinct fates during early development.

Existing Adult Stem Cells

As has been discussed in other chapters, numerous types of precursor cells have been isolated in adult tissues. Although these cells tend to be relatively rare and are dispersed throughout the tissues, they hold great potential for clinical application and tissue engineering. For example, tissues created using stem cells harvested from an adult patient could theoretically be used clinically in that patient without engendering an immune response. Moreover, the use of adult stem cells avoids the ethical concerns associated with the use of ES cells. In addition, adult-derived stem cells do not spontaneously differentiate as do ES cells, thus eliminating the formation of teratomas often seen with implantation of ES cells. The potential of adult stem cells for regenerative medicine is great; it is likely that these various cells will find clinical application in the upcoming decades.

Pluripotent Cell Lines are Tools for Future Research

Although the recent advances in reprogramming of adult somatic cells has generated a wave of interest in the scientific community, these cell lines will not likely replace hESC lines as tools for research and discovery. Rather, both categories of cells will find unique uses in the study of stem cell biology and the development and evaluation of therapeutic strategies. Pluripotent cells offer a number of potential clinical applications, especially for diseases with a genetic basis. However, researchers are just beginning to unlock the many factors that govern the cells' growth and differentiation. As scientists make strides toward understanding how these cells can be manipulated, additional applications, approaches, and techniques will likely emerge. As such, pluripotent cells will play a pivotal role in future research into the biology of development and the treatment of disease.

Bone Marrow

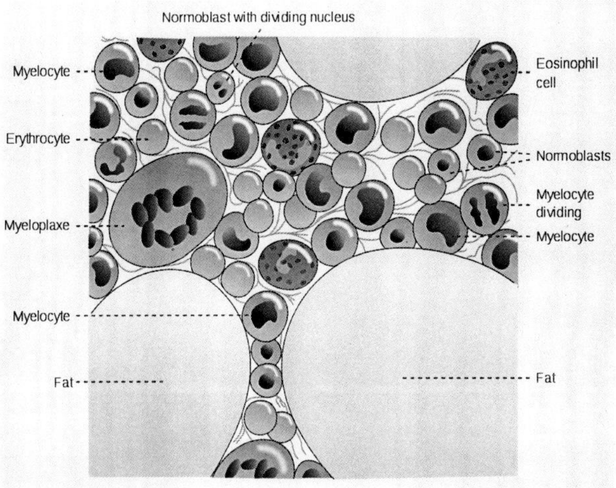

Figure: *Simplified illustration of cells in bone marrow.*

Bone marrow is the flexible tissue found in the interior of bones. In humans, bone marrow in large bones produces new blood cells. On average, bone marrow constitutes 4% of the total body mass of humans; in adults weighing 65 kg (143 lbs), bone marrow accounts for approximately 2.6 kg (5.7 lbs). The hematopoietic compartment of bone marrow produces approximately 500 billion blood cells per day, which use the bone marrow vasculature as a conduit to the body's systemic circulation.

Marrow Types

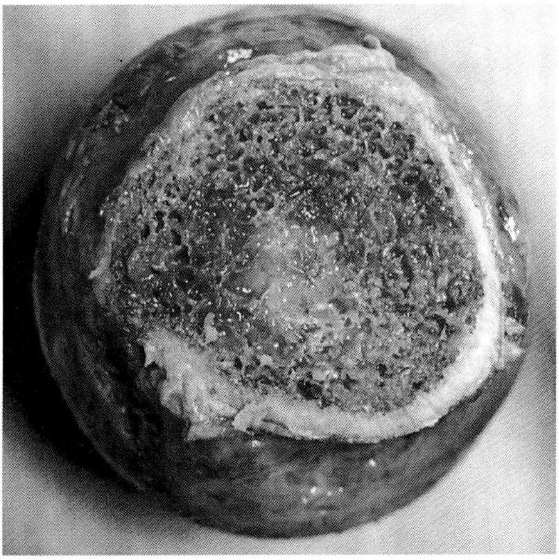

Figure: *A femur with a cortex of cortical bone and medulla of trabecular bone, showing its red bone marrow and a focus of yellow bone marrow.*

There are two types of bone marrow: medulla ossium rubra (*red marrow* - consisting mainly of hematopoietic tissue) and medulla ossium flava (*yellow marrow* - consisting mainly of fat cells). Red blood cells, platelets and most white blood cells arise in red marrow. Both types of bone marrow contain numerous blood vessels and capillaries. At birth, all bone marrow is red. With age, more and more of it is converted to the yellow type; only around half of adult bone marrow is red. Red marrow is found mainly in the flat bones, such as the hip bone, breast bone, cranium, ribs, vertebrae and shoulder blades, and in the cancellous ("spongy") material at the epiphyseal ends of long bones such as the femur and humerus. Yellow marrow is found in the medullary cavity, the hollow interior of the middle portion of long bones. In cases of severe blood loss, the body can convert yellow marrow back to red marrow to increase blood cell production.

Stroma

The stroma of the bone marrow is all tissue not directly involved in the primary function of hematopoiesis. Yellow bone marrow makes up the majority of bone marrow stroma, in addition to smaller concentrations of stromal cells located in the red bone marrow.

Though not as active as parenchymal red marrow, stroma is indirectly involved in hematopoiesis, since it provides the hematopoietic microenvironment that facilitates hematopoiesis by the parenchymal cells.

For instance, they generate colony stimulating factors, which have a significant effect on hematopoiesis.

Cells that constitute the bone marrow stroma are:

- fibroblasts (reticular connective tissue)
- macrophages
- adipocytes
- osteoblasts
- osteoclasts
- endothelial cells, which form the sinusoids. These derive from endothelial stem cells, which are also present in the bone marrow.

Macrophages contribute especially to red blood cell production, as they deliver iron for hemoglobin production.

Bone Marrow Barrier

The blood vessels of the bone marrow constitute a barrier, inhibiting immature blood cells from leaving the marrow. Only mature blood cells contain the membrane proteins required to attach to and pass the blood vessel endothelium.

Hematopoietic stem cells may also cross the bone marrow barrier, and may thus be harvested from blood.

Mesenchymal Stem Cells

The bone marrow stroma contain mesenchymal stem cells (MSCs), also called *marrow stromal cells*. These are multipotent stem cells that can differentiate into a variety of cell types. MSCs have been shown to differentiate, in vitro or in vivo, into osteoblasts, chondrocytes, myocytes, adipocytes and beta-pancreatic islets cells. MSCs can also transdifferentiate into neuronal cells.

Red Marrow Parenchyma

Types of Cells

Group	*Cell type*	*Average fraction*
Cells of myelopoiesis	Myeloblasts	0.9%
	Promyelocytes	3.3%
	Neutrophilic myelocytes	12.7%
	Eosinophilic myelocytes	0.8%
	Neutrophilic metamyelocytes	15.9%
	Eosinophilic metamyelocytes	1.2%
	Neutrophilic band cells	12.4%
	Eosinophilic band cells	0.9%
	Segmented neutrophils	7.4%
	Segmented eosinophils	0.5%
	Segmented basophils +	0.1%
Cells of erythropoiesis	Mast cells Pronormoblasts	0.6%
	Basophilic normoblasts	1.4%
	Polychromatic normoblasts	21.6%
	Orthochromatic normoblast	2.0%
Other cell types	Megakaryocytes	< 0.1%
	Plasma cells	1.3%
	Reticular cells	0.3%
	Lymphocytes	16.2%
	Monocytes	0.3%

In addition, the bone marrow contains hematopoietic stem cells, which give rise to the three classes of blood cells that are found in the circulation:

white blood cells (leukocytes), red blood cells (erythrocytes), and platelets (thrombocytes).

Compartmentalisation

There is biologic compartmentalisation in the bone marrow, in that certain cell types tend to aggregate in specific areas. For instance, erythrocytes, macrophages, and their precursors tend to gather around blood vessels, while granulocytes gather at the borders of the bone marrow.

Lymphatic Role

The red bone marrow is a key element of the lymphatic system, being one of the primary lymphoid organs that generate lymphocytes from immature hematopoietic progenitor cells. The bone marrow and thymus constitute the primary lymphoid tissues involved in the production and early selection of lymphocytes. Furthermore, bone marrow performs a valve-like function to prevent the backflow of lymphatic fluid in the lymphatic system.

Diseases Involving the Bone Marrow

The normal bone marrow architecture can be displaced by malignancies, aplastic anemia, or infections such as tuberculosis, leading to a decrease in the production of blood cells and blood platelets. In addition, cancers of the hematologic progenitor cells in the bone marrow can arise; these are the leukemias. Exposure to radiation or chemotherapy will kill many of the rapidly dividing cells of the bone marrow, and will therefore result in a depressed immune system. Many of the symptoms of radiation sickness are due to damage to the bone marrow cells. To diagnose diseases involving the bone marrow, a bone marrow aspiration is sometimes performed. This typically involves using a hollow needle to acquire a sample of red bone marrow from the crest of the ilium under general or local anesthesia.

Examination

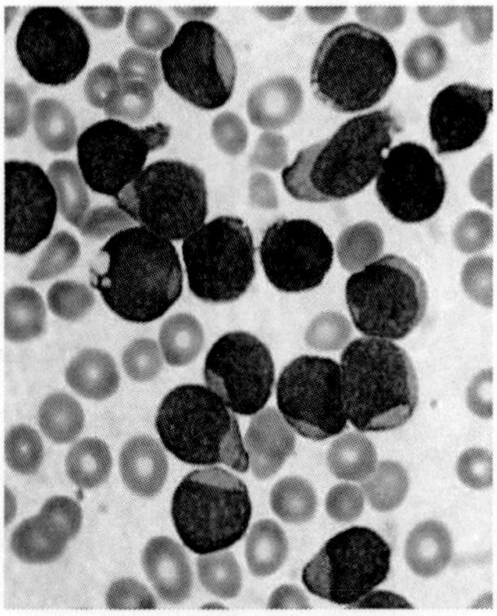

Figure: *A Wright's-stained bone marrow aspirate smear from a patient with leukemia.*

Bone marrow examination is the pathologic analysis of samples of bone marrow obtained by *bone marrow biopsy* and *bone marrow aspiration*. Bone marrow examination is used in the diagnosis of a number of conditions, including leukemia, multiple myeloma, anemia, and pancytopenia. The bone marrow produces the cellular elements of the blood, including platelets, red blood cells and white blood cells. While much information can be gleaned by testing the blood itself (drawn from a vein by phlebotomy), it is sometimes necessary to examine the source of the blood cells in the bone marrow to obtain more information on hematopoiesis; this is the role of bone marrow aspiration and biopsy.

Donation and Transplantation of Bone Marrow

In a bone marrow transplant, hematopoietic stem cells are removed from a person and infused into another person (allogenic) or into the same person at a later time (autologous). If donor and recipient are compatible, these infused cells will then travel to the bone marrow and initiate blood cell production. Transplantation from one person to another is performed in severe cases of disease of the bone marrow. The patient's marrow is first killed off with drugs or radiation, and then the new stem cells are introduced. Before radiation therapy or chemotherapy in cases of cancer, some of the patient's hematopoietic stem cells are sometimes harvested and later infused back when the therapy is finished to restore the immune system.

Harvesting

The stem cells are harvested directly from the red marrow in the crest of the ilium, often under general anesthesia. The procedure is minimally invasive and does not require stitches afterwards. Depending on the donor health and reaction to the procedure, the actual harvesting can be an outpatient procedure or requiring 1–2 days of recovery in the hospital. Another option is to administer certain drugs that stimulate the release of stem cells from the bone marrow into circulating blood. An IV is inserted into the donor's arm, and the stem cells are filtered out of the blood. The procedure is similar to donating blood or platelets. It may also be taken from the sternum. The tibia may seem a good source, since it is very superficial. However, except in children, this bone marrow does not contain any substantial amount of red bone marrow, only yellow bone marrow. In newborns, stem cells may be retrieved from the umbilical cord.

Bone Marrow as a Food

Many cultures have utilized bone marrow as a food throughout history. Some anthropologists believe that early humans were scavengers rather than hunters. Marrow would then have been a food source (largely for its fat content) for tool-using hominids, who were able to crack open the bones of carcasses left by apex predators such as lions.

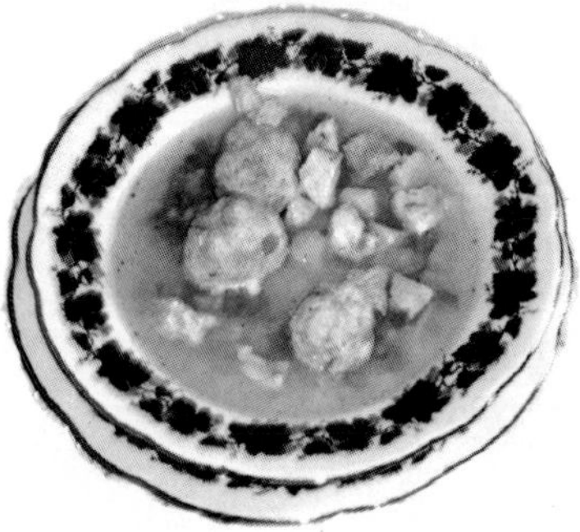

Figure: *In some parts of Germany, beef soup is served with "Markklößchen" (bone marrow balls).*

In Vietnam, beef bone marrow is used as the soup base for the national staple dish, *phß*, while in the Philippines, the soup *bulalo* is made primarily of beef stock and marrow bones, seasoned with vegetables and boiled meat; a similar soup in the Philippines is called *kansi*. In Indonesia, bone marrow is called *sumsum* and can be found especially in Minangkabau cuisine. *Sumsum* is often cooked as soup or as *gulai* (a curry-like dish). In India and Pakistan, slow-cooked marrow is the core ingredient of the dish *nalli nihari*.

In Hungary, tibia is a main ingredient of beef soup; the bone is chopped into short (10–15 cm) pieces and the ends are covered with salt to prevent the marrow from leaking from the bone while cooking. Upon serving the soup, the marrow is usually spread on toast.

Beef bone marrow is also the main ingredient in Italian dish ossobuco (braised veal shanks), and beef marrow bones are often included in the French pot-au-feu broth, the cooked marrow being traditionally eaten on toasted bread with sprinkled coarse sea salt.

In Iranian cuisine, lamb shanks are usually broken before cooking to allow diners to suck out and eat the marrow when the dish is served. Similar practices are also common in Pakistani cuisine. Some Native Alaskans eat the bone marrow of caribou and moose. European diners in the 18th century often used a marrow scoop (or marrow spoon), often of silver and with a long thin bowl, as a table implement for removing marrow from a bone. Though once used in various preparations, including pemmican, bone marrow's popularity is now relatively limited in the western world. Nonetheless, it remains in use in some gourmet restaurants and is popular among food enthusiasts.

Stem Cell Treatments and Veterinary Applications

Stem cell treatments are a type of intervention strategy that introduces new cells into damaged tissue in order to treat disease or injury. Many medical researchers believe that stem cell treatments have the potential to change the face of human disease and alleviate suffering.

The ability of stem cells to self-renew and give rise to subsequent generations with variable degrees of differentiation capacities, offers significant potential for generation of tissues that can potentially replace diseased and damaged areas in the body, with minimal risk of rejection and side effects. A number of stem cell therapies exist, but most are at experimental stages or costly, with the notable exception of bone marrow transplantation. Medical researchers anticipate that adult and embryonic stem cells will soon be able to treat cancer, Type 1 diabetes mellitus, Parkinson's disease, Huntington's disease, Celiac Disease, cardiac failure, muscle damage and neurological disorders, and many others. Nevertheless, before stem cell therapeutics can be applied in the clinical setting, more research is necessary to understand stem cell behaviour upon transplantation as well as the mechanisms of stem cell interaction with the diseased/injured microenvironment.

Current Treatments

For over 30 years, bone marrow, and more recently, umbilical cord blood stem cells, have been used to treat cancer patients with conditions such as leukemia and lymphoma. During chemotherapy, most growing cells are killed by the cytotoxic agents. These agents, however, cannot discriminate between the leukemia or neoplastic cells, and the hematopoietic stem cells within the bone marrow. It is this side effect of conventional chemotherapy strategies that the stem cell transplant attempts to reverse; a donor's healthy bone marrow reintroduces functional stem cells to replace the cells lost in the host's body during treatment.

Potential Treatments

Brain Damage: Stroke and traumatic brain injury lead to cell death, characterized by a loss of neurons and oligodendrocytes within the brain. Healthy adult brains contain neural stem cells which divide to maintain general stem cell numbers, or become progenitor cells. In healthy adult animals, progenitor cells migrate within the brain and function primarily to maintain neuron populations for olfaction (the sense of smell). Interestingly, in pregnancy and after injury, this system appears to be regulated by growth factors and can increase the rate at which new brain matter is formed. Although the reparative process appears to initiate following trauma to the brain, substantial recovery is rarely observed in adults, suggesting a lack of robustness. Stem cells may also be used to treat brain degeneration, such as in Parkinson's and Alzheimer's disease.

Cancer

The development of gene therapy strategies for treatment of intracranial tumours offers much promise, and has shown to be successful in the treatment of some dogs; although research in this area is still at an early stage.

Using conventional techniques, brain cancer is difficult to treat because it spreads so rapidly. Researchers at the Harvard Medical School transplanted human neural stem cells into the brain of rodents that received intracranial tumours. Within days, the cells migrated into the cancerous area and produced cytosine deaminase, an enzyme that converts a nontoxic prodrug into a chemotheraputic agent. As a result, the injected substance was able to reduce the tumor mass by 81 percent.

The stem cells neither differentiated nor turned tumorigenic. Some researchers believe that the key to finding a cure for cancer is to inhibit proliferation of cancer stem cells. Accordingly, current cancer treatments are designed to kill cancer cells. However, conventional chemotherapy treatments cannot discriminate between cancerous cells and others. Stem cell therapies may serve as potential treatments for cancer. Research on treating Lymphoma using adult stem cells is underway and has had human trials. Essentially, chemotherapy is used to completely destroy the patients own lymphocytes, and stem cells injected, eventually replacing the immune system of the patient with that of the healthy donor.

Spinal Cord Injury

A team of Korean researchers reported on November 25, 2003, that they had transplanted multipotent adult stem cells from umbilical cord blood to a patient suffering from a spinal cord injury and that following the procedure, she could walk on her own, without difficulty.

The patient had not been able to stand up for roughly 19 years. For the unprecedented clinical test, the scientists isolated adult stem cells from umbilical cord blood and then injected them into the damaged part of the spinal cord. According to the October 7, 2005 issue of *The Week*, University of California, Irvine researchers transplanted multipotent human fetal-derived neural stem cells into paralysed mice, resulting in locomotor improvements four months later. The observed recovery was associated with differentiation of transplanted cells into new neurons and oligodendrocytes- the latter of which forms the myelin sheath around axons of the central nervous system, thus insulating neural impulses and facilitating communication with the brain.

In January 2005, researchers at the University of Wisconsin–Madison differentiated human blastocyst stem cells into neural stem cells, then into premature motor neurons, and finally into spinal motor neurons, the cell type that, in the human body, transmits messages from the brain to the spinal cord and subsequently mediates motor function in the periphery. The newly generated motor neurons exhibited electrical activity, the signature action of neurons. Lead researcher Su-Chun Zhang described the process as "[teaching] the blastocyst stem cells to change step by step, where each step has different conditions and a strict window of time."

Transformation of blastocyst stem cells into motor neurons had eluded researchers for decades. While Zhang's findings were a significant contribution to the field, the ability of transplanted neural cells to establish communication with neighbouring cells remains unclear. Accordingly, studies using chicken embryos as a model organism can be an effective proof-of-concept experiment. If functional, the new cells could be used to treat diseases like Lou Gehrig's disease, muscular dystrophy, and spinal cord injuries.

Heart Damage

Several clinical trials targeting heart disease have shown that adult stem cell therapy is safe, effective, and equally efficient in treating old and recent infarcts. Adult stem cell therapy for treating heart disease was commercially available in at least five continents at the last count (2007).

Possible mechanisms of recovery include:

- Generation of heart muscle cells
- Stimulation of growth of new blood vessels to repopulate damaged heart tissue
- Secretion of growth factors
- Assistance via some other mechanism.

It may be possible to have adult bone marrow cells differentiate into heart muscle cells.

Haematopoiesis (Blood Cell Formation)

The specificity of the human immune cell repertoire is what allows the human body to defend itself from rapidly adapting antigens. However, the immune system is vulnerable to degradation upon the pathogenesis of disease, and because of the critical role that it plays in overall defence, its degradation is often fatal to the organism as a whole. Diseases of hematopoietic cells are called hematopathology. The specificity of the immune cells is what allows recognition of foreign antigens, causing further challenges in the treatment of immune disease.

Identical matches between donor and recipient must be made for successful transplantation treatments, but matches are uncommon, even between first-degree relatives. Research using both hematopoietic adult stem cells and embryonic stem cells has provided insight into the possible mechanisms and methods of treatment for many of these ailments. Fully mature human red blood cells may be generated *ex vivo* by hematopoietic stem cells (HSCs), which are precursors of red blood cells. In this process, HSCs are grown together with stromal cells, creating an environment that mimics the conditions of bone marrow, the natural site of red blood cell growth.

Erythropoietin, a growth factor, is added, coaxing the stem cells to complete terminal differentiation into red blood cells. Further research into this technique should have potential benefits to gene therapy, blood transfusion, and topical medicine.

Baldness

Hair follicles also contain stem cells, and some researchers predict research on these follicle stem cells may lead to successes in treating baldness through an activation of the stem cells progenitor cells. This treatment is expected to work by activating already existing stem cells on the scalp.

Later treatments may be able to simply signal follicle stem cells to give off chemical signals to nearby follicle cells which have shrunk during the aging process, which in turn respond to these signals by regenerating and once again making healthy hair. Most recently, Dr. Aeron Potter of the University of California has claimed that stem cell therapy led to a significant and visible improvement in follicular hair growth. Results from his experiments are under review by the journal *Science* (journal).

Missing Teeth

In 2004, scientists at King's College London discovered a way to cultivate a complete tooth in mice and were able to grow them stand-alone in the laboratory. Researchers are confident that this technology can be used to grow live teeth in human patients. In theory, stem cells taken from the patient could be coaxed in the lab into turning into a tooth bud which, when implanted in the gums, will give rise to a new tooth, and would be expected to grow within two months. It will fuse with the jawbone and release chemicals that encourage nerves and blood vessels to connect with it. The process is similar to what happens when humans grow their original adult teeth.

Many challenges remain, however, before stem cells could be a choice for the replacement of missing teeth in the future.

Deafness

Heller has reported success in regrowing cochlea hair cells with the use of embryonic stem cells.

Blindness and Vision Impairment

Since 2003, researchers have successfully transplanted corneal stem cells into damaged eyes to restore vision. "Sheets of retinal cells used by the team are harvested from aborted fetuses, which some people find objectionable." When these sheets are transplanted over the damaged cornea, the stem cells stimulate renewed repair, eventually restore vision. The latest such development was in June 2005, when researchers at the Queen Victoria Hospital of Sussex,

England were able to restore the sight of forty patients using the same technique. The group, led by Dr. Sheraz Daya, was able to successfully use adult stem cells obtained from the patient, a relative, or even a cadaver. Further rounds of trials are ongoing. In April 2005, doctors in the UK transplanted corneal stem cells from an organ donor to the cornea of Deborah Catlyn, a woman who was blinded in one eye when acid was thrown in her eye at a nightclub. The cornea, which is the transparent window of the eye, is a particularly suitable site for transplants. In fact, the first successful human transplant was a cornea transplant. The absence of blood vessels within the cornea makes this area a relatively easy target for transplantation. The majority of corneal transplants carried out today are due to a degenerative disease called keratoconus.

The University Hospital of New Jersey reports that the success rate for growth of new cells from transplanted stem cells varies from 25 percent to 70 percent. In 2009, researchers at the University of Pittsburgh Medical Centre demonstrated that stem cells collected from human corneas can restore transparency without provoking a rejection response in mice with corneal damage.

Amyotrophic Lateral Sclerosis (ALS)

Stem cells have resulted in significant locomotor improvements in rats with an Amyotrophic lateral sclerosis-like disease. In a rodent model that closely mimics the human form of ALS, animals were injected with a virus to kill the spinal cord motor nerves which mediate movement. Animals subsequently received stem cells in the spinal cord. Transplanted cells migrated to the sites of injury, contributed to regeneration of the ablated nerve cells, and restored locomotor function.

Graft Vs Host Disease and Crohn's Disease

Phase III clinical trials expected to end in second-quarter 2008 were conducted by Osiris Therapeutics using their in-development product Prochymal, derived from adult bone marrow. The target disorders of this therapeutic are graft-versus-host disease and Crohn's disease.

Neural and Behavioural Birth Defects

A team of researchers led by Prof. Joseph Yanai were able to reverse learning deficits in the offspring of pregnant mice who were exposed to heroin and the pesticide organophosphate. This was done by direct neural stem cell transplantation into the brains of the offspring. The recovery was almost 100 percent, as shown both in behavioural tests and objective brain chemistry tests. Behavioural tests and learning scores of the treated mice showed rapid improvement after treatment, providing results that rivalled non-treated mice.

On the molecular level, brain chemistry of the treated animals was also restored to normal. Through the work, which was supported by the US National Institutes of Health, the US-Israel Binational Science Foundation and the Israel anti-drug authorities, the researchers discovered that the stem cells worked even in cases where most of the cells died out in the host brain.

The scientists found that before they die the neural stem cells succeed in inducing the host brain to produce large numbers of stem cells which repair the damage. These findings, which answered a major question in the stem cell research community, were published in January 2008 in the leading journal, *Molecular Psychiatry*. Scientists are now developing procedures to administer the neural stem cells in the least invasive way possible - probably via blood vessels, making therapy practical and clinically feasible. Researchers also plan to work on developing methods to take cells from the patient's own body, turn them into stem cells, and then transplant them back into the patient's blood via the blood stream. Aside from decreasing the chances of immunological rejection, the approach will also eliminate the controversial ethical issues involved in the use of stem cells from human embryos.

Diabetes

Diabetes patients lose the function of insulin-producing beta cells within the pancreas. Human embryonic stem cells may be grown in cell culture and stimulated to form insulin-producing cells that can be transplanted into the patient.

However, clinical success is highly dependent on the development of the following procedures:

- Transplanted cells should proliferate
- Transplanted cells should differentiate in a site-specific manner
- Transplanted cells should survive in the recipient (prevention of transplant rejection)
- Transplanted cells should integrate within the targeted tissue
- Transplanted cells should integrate into the host circuitry and restore function.

Orthopaedics

Clinical case reports in the treatment of orthopaedic conditions have been reported. To date, the focus in the literature for musculoskeletal care appears to be on mesenchymal stem cells. Centeno et al. have published MRI evidence of increased cartilage and meniscus volume in individual human subjects. The results of trials that include a large number of subjects, are yet to be published. However, a published safety study conducted in a group of 227 patients over a 3-4 year period shows adequate safety and minimal complications associated with mesenchymal cell transplantation.

Wakitani has also published a small case series of nine defects in five knees involving surgical transplantation of mesenchymal stem cells with coverage of the treated chondral defects.

Wound Healing

Stem cells can also be used to stimulate the growth of human tissues. In an adult, wounded tissue is most often replaced by scar tissue, which is characterised in the skin by disorganised collagen structure, loss of hair follicles and irregular vascular structure. In the case of wounded fetal tissue, however, wounded tissue is replaced with normal tissue through the activity of stem cells. A possible method for tissue regeneration in adults is to place adult stem cell "seeds" inside a tissue bed "soil" in a wound bed and allow the stem cells to stimulate differentiation in the tissue bed cells. This method elicits a regenerative response more similar to fetal wound-healing than adult scar tissue formation. Researchers are still investigating different aspects of the "soil" tissue that are conducive to regeneration.

Infertility

Culture of human embryonic stem cells in mitotically inactivated porcine ovarian fibroblasts (POF) causes differentiation into germ cells (precursor cells of oocytes and spermatozoa), as evidenced by gene expression analysis. Human embryonic stem cells have been stimulated to form Spermatozoon-like cells, yet still slightly damaged or malformed. It could potentially treat azoospermia.

Clinical Trials

On January 23, 2009, the US Food and Drug Administration gave clearance to Geron Corporation for the initiation of the first clinical trial of an embryonic stem cell-based therapy on humans. The trial will evaluate the drug GRNOPC1, embryonic stem cell-derived oligodendrocyte progenitor cells, on patients with acute spinal cord injury. As of mid-2010 hundreds of phase III clinical trials involving stem cells have been registered.

Potential Contributions to Veterinary Medicine

Research currently conducted on horses, dogs, and cats can benefit the development of stem-cell treatments in veterinary medicine and can target a wide range of injuries and diseases such as myocardial infarction, stroke, tendon and ligament damage, osteoarthritis, osteochondrosis and muscular dystrophy both in large animals, as well as humans. While investigation of cell-based therapeutics generally reflects human medical needs, the high degree of frequency and severity of certain injuries in racehorses has put veterinary medicine at the forefront of this novel regenerative approach.

Companion animals can serve as clinically relevant models that closely mimic human disease.

Development of Regenerative Treatment Models

Veterinary applications of stem cell therapy as a means of tissue regeneration have been largely shaped by research that began with the use of adult-derived mesenchymal stem cells to treat animals with injuries or defects affecting bone, cartilage, ligaments and/or tendons. Because mesenchymal stem cells can differentiate into the cells that make up bone, cartilage, tendons, and ligaments (as well as muscle, fat, and possibly other tissues), they have been the main type of stem cells studied in the treatment of diseases affecting these tissues. Mesenchymal stem cells are primarily derived from adipose tissue or bone marrow.

Since an elevated immune response following cell transplantation may result in rejection of exogenous cells (except in the case of cells derived from a very closely genetically related individual), mesenchymal stem cells are often derived from the patient prior to injection in a process known as autologous transplantation. Surgical repair of bone fractures in dogs and sheep has demonstrated that engraftment of mesenchymal stem cells derived from a genetically different donor within the same species, termed allogeneic transplantation, does not elicit an immunological response in the recipient animal and can mediate regeneration of bone tissue in major bony fractures and defects. Stem cells can speed up bone repair in fractures/defects that would normally require extensive grafting, suggesting that mesenchymal stem cell use may provide a useful alternative to conventional grafting techniques. Treating tendon and ligament injuries in horses using stem cells, whether derived from adipose tissue or bone-marrow, has support in the veterinary literature.

While further studies are necessary to fully characterize the use of cell-based therapeutics for treatment of bone fractures, stem cells are thought to mediate repair via five primary mechanisms: 1) providing an anti-inflammatory effect, 2) homing to damaged tissues and recruiting other cells, such as endothelial progenitor cells, that are necessary for tissue growth, 3) supporting tissue remodelling over scar formation, 4) inhibiting apoptosis, and 5) differentiating into bone, cartilage, tendon, and ligament tissue.

Significance of Stem Cell Microenvironments

The microenvironment into which stem cells are transplanted significantly alters the capacity of engrafted cells for recovery and repair. The microenviroment provides growth factors and other chemical signals that guide appropriate differentiation of transplanted cell populations and direct transplanted cells to sites of trauma or disease. Repair and recovery can then be mediated via three primary mechanisms: 1) formation and/or recruitment of new blood cells to the damaged region; 2) prevention of programed cell death or apoptosis; and 3) suppression of inflammation. To further enrich blood supply to the damaged areas, and consequently promote tissue regeneration, platelet-rich plasma could be used in conjunction with stem cell transplantation. The efficacy of some stem cell populations may also be affected by the method of delivery; for instance, to regenerate bone, stem cells are often introduced in a scaffold where they produce the minerals necessary for generation of functional bone.

Sources of Autologous (Patient-derived) Stem Cells

Autologous stem cells intended for regenerative therapy are generally isolated either from the patient's bone marrow or from adipose tissue. The number of stem cells transplanted into damaged tissue may alter efficacy of treatment. Accordingly, stem cells derived from bone marrow aspirates, for instance, are cultured in specialised laboratories for expansion to millions of cells. Although adipose-derived tissue also requires processing prior to use, the culturing methodology for adipose-derived stem cells is not as extensive as that for bone marrow-derived cells. While it is thought that bone-marrow derived stem cells are preferred for bone, cartilage, ligament, and tendon repair, others believe that the less challenging collection techniques and the multicellular microenvironment already present in adipose-derived stem cell fractions make the latter the preferred source for autologous transplantation.

Currently Available Treatments for Horses and Dogs Suffering From Orthopaedic Conditions

Autologous or allogeneic stem cells are currently used as an adjunctive therapy in the surgical repair of some types of fractures in dogs and horses. Autologous stem cell-based treatments for ligament injury, tendon injury, osteoarthritis, osteochondrosis, and sub-chondral bone cysts have been commercially available to practicing veterinarians to treat horses since 2003 in the United States and since 2006 in the United Kingdom. Autologous stem-cell based treatments for tendon injury, ligament injury, and osteoarthritis in dogs have been available to veterinarians in the United States since 2005. Over 3000 privately-owned horses and dogs have been treated with autologous adipose-derived stem cells. The efficacy of these treatments has been shown in double-blind clinical trials for dogs with osteoarthritis of the hip and elbow and horses with tendon damage. The efficacy of using stem cells, whether adipose-derived or bone-marrow derived, for treating tendon and ligament injuries in horses has support in the veterinary literature.

Developments in Stem Cell Treatments in Veterinary Internal Medicine

Currently, research is being conducted to develop stem cell treatments for: 1) horses suffering from COPD, neurologic disease, and laminitis; and 2) dogs and cats suffering from heart disease, liver disease, kidney disease, neurologic disease, and immune-mediated disorders.

Embryonic Stem Cell Controversy

There is widespread controversy over the use of human embryonic stem cells. This controversy primarily targets the techniques used to derive new embryonic stem cell lines, which often requires the destruction of the blastocyst. Opposition to the use of human embryonic stem cells in research is often based on philosophical, moral or religious objections.

Stem Cell Treatments Around the World

China

Stem cell research and treatment is currently being practiced at a clinical level in the People's Republic of China. The Ministry of Health of the People's Republic of China has permitted the use of stem cell therapy for conditions beyond those approved of in Western countries such as the United States, United Kingdom, and Australia.

The Western World has scrutinized China for its failed attempts to meet international documentation standards of these trials and procedures, despite the overwhelmingly positive anecdotal results. Stem cell therapies provided in China utilize a variety of cell types including umbilical cord stem cells and olfactory ensheathing cells. The stem cells are then expanded in centralized blood banks before being used in stem cell treatments. State-funded companies based in the Shenzhen Hi-Tech Industrial Zone treat the symptoms of numerous disorders with adult stem cell therapy. Hospitals throughout eastern China provide numerous therapies to patients in coordination with the stem cell providers. These companies' therapies are currently focused on the treatment of neurodegenerative and cardiovascular disorders.

The most radical successes of Chinese adult stem cell therapy have been in treating the brain. These therapies administer stem cells directly to the brain to promote greater motor and brain function in patients with Cerebral Palsy, Alzheimer's, and brain injuries. However, retrospective studies have shown that Chinese use of fetal-derived brain tissue in spinal cord injured human subjects were not as promising as once thought: the phenotype and the fate of the transplanted cells, described as olfactory ensheathing cells, were unknown. As well, perioperative morbidity and lack of functional benefit were identified as the most serious clinical shortcomings. Furthermore, the extent of regulatory policy in the use of stem cell therapies in China is unclear. In the absence of a valid clinical trials protocol, and more regulatory oversight, Western regulatory agencies advise patients and physicians to be cautious when selecting Chinese stem cell therapeutic centres.

Mexico

Stem cell treatment is currently being practised at a clinical level in Mexico. An International Health Department Permit (COFEPRIS) is required. This permit allows the use of stem cell types beyond those approved of in Western countries such as the United States or Europe. Stem cell therapies provided in Mexico utilize patient Adipose, Bone Marrow, or Donor Placenta sources.

South Korea

In 2005, South Korean scientists claimed to have generated stem cells that were tailored to match the recipient. Each of the 11 new stem cell lines was developed using somatic cell nuclear transfer (SCNT) technology. The resultant cells were thought to match the genetic material of the recipient, thus suggesting minimal to no cell rejection. This study, however, was eventually discredited as the primary researcher, Dr. Woo Suk Hwang, admitted to using cells obtained from his research staff. In December 2005, claims were put forward that his research had been manipulated to wrongfully indicate positive results. Later that month, these claims were confirmed by an academic panel.

Ukraine

Today, Ukraine is permitted to perform clinical trials of stem cell treatments (Order of the MH of Ukraine 630 "About carrying out clinical trials of stem cells", 2008) for the treatment of these pathologies: pancreatic necrosis, cirrhosis, hepatitis, burn disease, diabetes, multiple sclerosis, critical lower limb ischemia. The first medical institution granted the right to conduct clinical trials became the "Institute of Cell Therapy"(Kiev).

Cell Potency

The potency of a cell specifies its differentiation potential, or potential to differentiate into different cell types. Totipotency is the ability of a single cell to divide and produce all the differentiated cells in an organism, including extraembryonic tissues. Totipotent cells include spores and zygotes. In some organisms, cells can dedifferentiate and regain totipotency. For example, a plant cutting or callus can be used to grow an entire plant. Human development begins when a sperm fertilizes an egg and creates a single totipotent cell (zygote). In the first hours after fertilization, this cell divides into identical totipotent cells, which can later develop into any of the three germ layers of a human (endoderm, mesoderm, or ectoderm) and into cells of the cytotrophoblast layer or syncytiotrophoblast layer of the placenta. After reaching the 16-cell stage, the totipotent cells of the morula differentiate into cells that will eventually become either the Blastocyst's Inner cell mass or outer trophoblasts. Approximately four days after fertilization and after several cycles of cell division, these totipotent cells begin to specialize. The inner cell mass, the source of embryonic stem cells, is pluripotent, not totipotent. Research on *Caenorhabditis elegans* suggests that multiple mechanisms including RNA regulation may play a role in maintaining totipotency at different stages of development in some species.

Pluripotency

In cell biology, pluripotency (from the Latin "plurimus" or "very many" and "potentia" or "powered") refers to a stem cell that has the potential to differentiate into any of the three germ layers: endoderm (interior stomach lining, gastrointestinal tract, the lungs), mesoderm (muscle, bone, blood, urogenital), or ectoderm (epidermal tissues and nervous system).

Pluripotent stem cells can give rise to any fetal or adult cell type. However, alone they cannot develop into a fetal or adult animal because they lack the potential to contribute to extraembryonic tissue, such as the placenta.

Progenitor Cell

A progenitor cell is a biological cell that, like a stem cell, has a tendency to differentiate into a specific type of cell, but is already more specific than a stem cell and is pushed to differentiate into its "target" cell. The most important difference between stem cells and progenitor cells is that stem cells can replicate indefinitely, whereas progenitor cells can only divide a limited number of times. Controversy about the exact definition remains and the concept is still evolving. The terms "progenitor cell" and "stem cell" are sometimes equated.

Properties

Most progenitors are described as oligopotent. In this point of view, they may be compared to adult stem cells. But progenitors are said to be in a further stage of cell differentiation. They are in the "centre" between stem cells and fully differentiated cells.

The kind of potency they have depends on the type of their "parent" stem cell and also on their niche. Like stem cells, mostly, they are formed and transported in a colony, with the right conditions for them to grow and differentiate into their target tissues. In contrast, some progenitor cells were found during research, and were isolated. After their marker was found, it was proven that these progenitor could move through the body and migrate towards the tissue where they are needed. Many properties are shared by adult stem cells and progenitor cells. But still, controversy remains because embryonic stem cells are true stem cells in that they are pluripotent and show unlimited capacity for self-renewal. In contrast, many cells termed adult stem cells would be better defined as progenitor cells, as their capacities for unlimited self-renewal and plasticity have not been comprehensively demonstrated.

Function

The majority of progenitor cells lie dormant or possess little activity in the tissue in which they reside. They exhibit slow growth and their main role is to replace cells lost by normal attrition. In case of tissue injury, damaged or dead cells, progenitor cells can be activated. Growth factors or cytokines are two substances which trigger the progenitors to mobilize towards the damaged tissue. At the same time, they start to differentiate into the target cells. Not all progenitors are mobile and are situated near the tissue of their target differentiation. When the cytokines, growth factors and other cell division enhancing stimulators take on the progenitors, a higher rate of cell division is introduced. It leads to the recovery of the tissue.

Examples

The characterisation or the defining principle of progenitor cells, in order to separate them from others, is based on the different cell markers rather than their morphological appearance.

- Satellite cells found in muscles. They play a major role in muscle cell differentiation and injury recoveries.
- Intermediate progenitor cells formed in the subventricular zone. Some of these transit amplifying neural progenitors migrate via rostral migratory stream to the olfactory bulb and differentiate further into specific types of neural cells.
- Bone marrow stromal cells, basal cell of epidermis have 10% of progenitor cell, although they are often classed as stem cells due to their high plasticity and potentially unlimited capacity for self renewal.
- Periosteum contains progenitor cells that develop into osteoblasts and chondroblasts.
- Pancreatic progenitor cells are among the most studied progenitors. They are used in research to develop a cure against diabetes type-1.
- Angioblasts or Endothelial progenitor cells (EPC). These are very important for research on fracture and wounds healing.

Development of the Human Cerebral Cortices

Before E40 (embryonic day 40), progenitor cells generate other progenitor cells; after that period progenitor cells produce only dissimilar daughters. The cells from a single progenitor cell form a proliferative unit that creates on cortical column; these columns contain a variety of neurons with different shapes.

Oligopotency

In biology, oligopotency is the ability of progenitor cells to differentiate into a few cell types. It is a degree of potency. Examples of oligopotent stem cells are the lymphoid or myeloid stem cells. Examples of progenitor cells are Vascular Stem Cells which have the capacity to become both Endothelial or smooth muscle cells.

Unipotency

In cell biology, a unipotent cell is one that has the capacity to develop/ differentiate into only one type of tissue/cell type. The most common of these in humans are skin cells. This cell has a unique property: self-renewal. This property distinguishes it from most other terminally differentiated non-stem cells. Hepatocytes, which constitute most of the Cytoplasmic mass of the human liver, are unipotent. The liver's ability to regenerate from as little as 25% of its original mass is attributed to this property. A close synonym for *unipotent cell* is *precursor cell.*

2

Cloning in Stem Cell Research

Somatic cell nuclear transfer, known as SCNT, can also be used to create embryos for research or therapeutic purposes. The most likely purpose for this is to produce embryos for use in stem cell research. This process is also called "research cloning" or "therapeutic cloning." The goal is not to create cloned human beings (called "reproductive cloning"), but rather to harvest stem cells that can be used to study human development and to potentially treat disease. While a clonal human blastocyst has been created, stem cell lines are yet to be isolated from a clonal source.

Therapeutic cloning is achieved by creating embryonic stem cells in the hopes of treating diseases such as diabetes and Alzheimer's. The process begins by taking out the nucleus that contains the DNA from an egg and putting it in a nucleus from an adult 52. In the case of someone with Alzheimer's disease, the nucleus from a skin cell of that patient is placed into an empty egg. The reprogrammed cell begins to develop into an embryo because the egg reacts with the transferred nucleus. The embryo will become genetically identical to the patient 52. The embryo will then form a blastocyst which has the potential to form become any cell in the body 53.

The reason why SCNT is used for cloning is because somatic cells can be easily acquired and cultured in the lab. This process can either add or delete specific genomes of farm animals. A key point to remember is that cloning is achieved when the oocyte maintains its normal functions and instead of using sperm and egg genomes to replicate, the oocyte is inserted into the donor's somatic cell nucleus 48. The oocyte will react on the somatic cell nucleus, the same way it would on sperm cells 48.

SCNT Process: The process of cloning a particular farm animal using SCNT is relatively the same for all animals. The first step is to collect the somatic cells from the animal that will be cloned. The somatic cells could be used immediately or stored in the laboratory for later use 48. The hardest part of SCNT is removing maternal DNA from an oocyte at metaphase II.

Once this has been done, the somatic nucleus can be inserted into an egg cytoplasm 48. This creates a one-cell embryo. The grouped somatic cell and egg cytoplasm are then introduced to an electrical current 48. This energy will hopefully allow the cloned embryo to begin development. The successfully developed embryos are then placed in surrogate recipients, such as a cow or sheep in the case of farm animals 48. Applications of SCNT SCNT is seen to be a great method for producing agriculture animals for food consumption. It successfully cloned sheep, cattle, goats, and pigs. Another benefit is SCNT is seen as a solution to clone endangered species that are on the verge of going extinct 48.

Limitations of SCNT: Along with the benefits, there are several limitations for the use of SCNT to clone. It was shown to be an inadequate form of cloning. An example of this is Dolly the sheep was the only successful clone from a group of 277 cloned embryos 47. There is also a high count of embryos dying midway through pregnancy 48. If the embryo makes it through, majority of the time the offspring are larger than normal and die quickly 48. There is also the risk of these cloned animals being born with severe birth defects 49. Not only are there complications for the off spring, but the surrogate mother can experience suffering as well 48.

Organism Cloning

Organism cloning (also called reproductive cloning) refers to the procedure of creating a new multicellular organism, genetically identical to another. In essence this form of cloning is an asexual method of reproduction, where fertilization or inter-gamete contact does not take place. Asexual reproduction is a naturally occurring phenomenon in many species, including most plants and some insects. Scientists have made some major achievements with cloning, including the asexual reproduction of sheep and cows. There is a lot of ethical debate over whether or not cloning should be used. However, cloning, or asexual propagation, has been common practice in the horticultural world for hundreds of years.

Horticultural

The term *clone* is used in horticulture to refer to descendants of a single plant which were produced by vegetative reproduction or apomixis. Many horticultural plant cultivars are clones, having been derived from a single individual, multiplied by some process other than sexual reproduction. As an example, some European cultivars of grapes represent clones that have been propagated for over two millennia. Other examples are potato and banana. Grafting can be regarded as cloning, since all the shoots and branches coming from the graft are genetically a clone of a single individual, but this particular kind of cloning has not come under ethical scrutiny and is generally treated as an entirely different kind of operation.

Many trees, shrubs, vines, ferns and other herbaceous perennials form clonal colonies. Parts of a large clonal colony often become detached from the parent, termed fragmentation, to form separate individuals. Some plants e.g. dandelion also form seeds asexually, termed apomixis, resulting in clonal populations of genetically identical individuals.

Parthenogenesis

Clonal derivation exists in nature in some animal species and is referred to as parthenogenesis (reproduction of an organism by itself without a mate). This is an asexual form of reproduction that is only found in females of some insects, crustaceans and lizards. The growth and development occurs without fertilization by a male. In plants, parthenogenesis means the development of an embryo from an unfertilized egg cell, and is a component process of apomixis. In species that use the XY sex-determination system, the offspring will always be female. An example is the "Little Fire Ant" (*Wasmannia auropunctata*), which is native to Central and South America but has spread throughout many tropical environments.

Artificial Cloning of Organisms

Artificial cloning of organisms may also be called *reproductive cloning.*

Methods

Reproductive cloning generally uses "somatic cell nuclear transfer" (SCNT) to create animals that are genetically identical. This process entails the transfer of a nucleus from a donor adult cell (somatic cell) to an egg that has no nucleus. If the egg begins to divide normally it is transferred into the uterus of the surrogate mother. Such clones are not strictly identical since the somatic cells may contain mutations in their nuclear DNA. Additionally, the mitochondria in the cytoplasm also contains DNA and during SCNT this DNA is wholly from the donor egg, thus the mitochondrial genome is not the same as that of the nucleus donor cell from which it was produced. This may have important implications for cross-species nuclear transfer in which nuclear-mitochondrial incompatibilities may lead to death.

Artificial *embryo splitting* or *embryo twinning* may also be used as a method of cloning, where an embryo is split in the maturation before embryo transfer. It is optimally performed at the 6- to 8-cell stage, where it can be used as an expansion of IVF to increase the number of available embryos. If both embryos are successful, it gives rise to monozygotic (identical) twins.

Dolly the Sheep

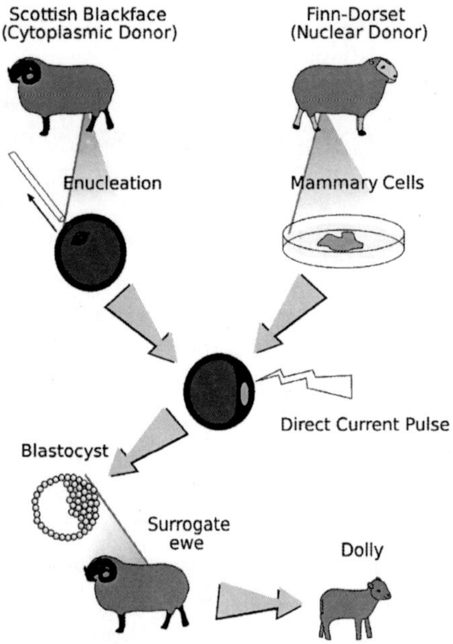

Dolly, a Finn-Dorset ewe, was the first mammal to have been successfully cloned from an adult cell. Dolly was formed by taking a cell from the udder of her biological mother. Her embryo was created by taking the cell and inserting it into a sheep ovum. The embryo was then placed inside a female sheep that went through a normal pregnancy 47. She was cloned at the Roslin Institute in Scotland and lived there from her birth in 1996 until her death in 2003 when she was six. Her stuffed remains were placed at Edinburgh's Royal Museum, part of the National Museums of Scotland.

Dolly was publicly significant because the effort showed that the genetic material from a specific adult cell, programmed to express only a distinct subset of its genes, can be reprogrammed to grow an entirely new organism. Before this demonstration, it had been shown by John Gurdon that nuclei from differentiated cells could give rise to an entire organism after transplantation into an enucleated egg. However, this concept was not yet demonstrated in a mamallian system. Cloning Dolly the sheep had a low success rate per fertilized egg; she was born after 237 eggs were used to create 29 embryos, which only produced three lambs at birth, only one of which lived. Seventy calves have been created and one third of them died young; Prometea took 277 attempts. Notably, although the first clones were frogs, no adult cloned frog has yet been produced from a somatic adult nucleus donor cell. There were early claims that Dolly the Sheep had pathologies resembling accelerated aging. Scientists speculated that Dolly's death in 2003 was related to the shortening of telomeres, DNA-protein complexes that protect the end of linear chromosomes. However, other researchers, including Ian Wilmut who led the team that successfully cloned Dolly, argue that Dolly's early death due to respiratory infection was unrelated to deficiencies with the cloning process. Many people knew Dolly as a failed scientific experiment where scientists tried to control the balance of nature. Her death created awareness and allowed many people to question the ethical limits regarding cloning 47. People are mostly concerned about the risks that cloned foods may have on human health and the environment (Jesper Lassen, 2005). They fear that the applications used to create Dolly will someday be used to clone humans 47.

Water Buffalo

On September 15, 2007, the Philippines announced its development of Southeast Asia's first cloned water buffalo. The Philippine Council for Agriculture, Forestry and Natural Resources Research and Development (PCARRD), under the Department of Science and Technology in Los Baños, Laguna approved this project.

Species Cloned

The modern cloning techniques involving nuclear transfer have been successfully performed on several species. Landmark experiments in chronological order:

- Tadpole: (1952) Many scientists questioned whether cloning had actually occurred and unpublished experiments by other labs were not able to reproduce the reported results.
- Carp: (1963) In China, embryologist Tong Dizhou produced the world's first cloned fish by inserting the DNA from a cell of a male carp into an egg from a female carp. He published the findings in a Chinese science journal.
- Mice: (1986) A mouse was the first mammal successfully cloned from an early embryonic cell. Soviet scientists Chaylakhyan, Veprencev, Sviridova, and Nikitin had the mouse "Masha" cloned. Research was published in the magazine "Biofizika" volume ÕÕÕII, issue 5 of 1987.
- Sheep: (1996) From early embryonic cells by Steen Willadsen. Megan and Morag cloned from differentiated embryonic cells in June 1995 and Dolly the sheep from a somatic cell in 1997.
- Rhesus Monkey: Tetra (January 2000) from embryo splitting
- Gaur: (2001) was the first endangered species cloned.

- Cattle: Alpha and Beta (males, 2001) and (2005) Brazil
- Cat: CopyCat "CC" (female, late 2001), Little Nicky, 2004, was the first cat cloned for commercial reasons
- Dog: Snuppy, a male Afghan hound was the first cloned dog (2005).
- Rat: Ralph, the first cloned rat (2003)
- Mule: Idaho Gem, a john mule born 4 May 2003, was the first horse-family clone.
- Horse: Prometea, a Haflinger female born 28 May 2003, was the first horse clone.
- Water Buffalo: Samrupa was the first cloned water buffalo. It was born on February 6, 2009, at India's Karnal National Diary Research Institute but died five days later due to lung infection.
- Pyrenean Ibex (2009) was the first extinct animal (extinct 2000) to be cloned back to life; the clone lived for seven minutes before dying of lung defects.
- Camel: (2009) Injaz, is the first cloned camel.

Human Cloning

Human cloning is the creation of a genetically identical copy of an existing or previously existing human. The term is generally used to refer to *artificial* human cloning; human clones in the form of identical twins are commonplace, with their cloning occurring during the natural process of reproduction. There are two commonly discussed types of human cloning: therapeutic cloning and reproductive cloning. Therapeutic cloning involves cloning adult cells for use in medicine and is an active area of research. Reproductive cloning would involve making cloned humans. A third type of cloning called replacement cloning is a theoretical possibility, and would be a combination of therapeutic and reproductive cloning. Replacement cloning would entail the replacement of an extensively damaged, failed, or failing body through cloning followed by whole or partial brain transplant.

The various forms of human cloning are controversial. There have been numerous demands for all progress in the human cloning field to be halted. Most scientific, governmental and religious organisations oppose reproductive cloning. The American Association for the Advancement of Science (AAAS) and other scientific organizations have made public statements suggesting that human reproductive cloning be banned until safety issues are resolved. Serious ethical concerns have been raised by the future possibility of harvesting organs from clones 50. Some people have considered the idea of growing organs separately from a human organism - in doing this, a new organ supply could be established without the moral implications of harvesting them from humans. Research is also being done on the idea of growing organs that are biologically acceptable to the human body inside of other organisms, such as pigs or cows, then transplanting them to humans, a form of xenotransplantation.

The first hybrid human clone was created in November 1998, by Advanced Cell Technologies. It was created from a man's leg cell, and a cow's egg whose DNA was removed. It was destroyed after 12 days. Since a normal embryo implants at 14 days, Dr Robert Lanza, ACT's director of tissue engineering, told the *Daily Mail* newspaper that the embryo could not be seen as a person before 14 days. While making an embryo, which may have resulted in a complete human had it been allowed to come to term, according to ACT: "[ACT's] aim was 'therapeutic cloning' not 'reproductive cloning'"

On January, 2008, Wood and Andrew French, Stemagen's chief scientific officer in California, announced that they successfully created the first 5 mature human embryos using DNA from adult skin cells, aiming to provide a source of viable embryonic stem cells. Dr. Samuel Wood and a colleague donated skin cells, and DNA from those cells was transferred to human eggs. It is not clear if the embryos produced would have been capable of further development, but Dr. Wood stated that if that were possible, using the technology for reproductive cloning would be both unethical and illegal. The 5 cloned embryos, created in Stemagen Corporation lab, in La Jolla, were destroyed.

Ethical Issues of Cloning

Because of recent technological advancements, the cloning of animals (and potentially humans) has been an issue. The Catholic Church and many religious organisations oppose all forms of cloning, on the grounds that life begins at conception. Judaism does not equate life with conception and, though some question the wisdom of cloning, Orthodox rabbis generally find no firm reason in Jewish law and ethics to object to cloning. From the standpoint of classical liberalism, concerns also exist regarding the protection of the identity of the individual and the right to protect one's genetic identity. Gregory Stock is a scientist and outspoken critic against restrictions on cloning research. Bioethicist Gregory Pence also attacks the idea of criminalizing attempts to clone humans.

The social implications of an artificial human production scheme were famously explored in Aldous Huxley's novel *Brave New World*.

On December 28, 2006, the U.S. Food and Drug Administration (FDA) approved the consumption of meat and other products from cloned animals. Cloned-animal products were said to be virtually indistinguishable from the non-cloned animals. Furthermore, companies would not be required to provide labels informing the consumer that the meat comes from a cloned animal. Critics have raised objections to the FDA's approval of cloned-animal products for human consumption, arguing that the FDA's research was inadequate, inappropriately limited, and of questionable scientific validity. Several consumer-advocate groups are working to encourage a tracking programme that would allow consumers to become more aware of cloned-animal products within their food.

Joseph Mendelson, legal director of the Centre for Food Safety, said that cloned food still should be labelled since safety and ethical issues about it remain questionable. Carol Tucker Foreman, director of food policy at the Consumer Federation of America, stated that FDA does not consider the fact that the results of some studies revealed that cloned animals have increased rates of mortality and deformity at birth. A major ethical concern people have is the simple fact that manipulating animals means altering the God-given nature of things. Another concern is that the biotechnologies used on animals may someday be used on humans 47. Some people may be more open to the idea of cloning of animals because most western countries have passed legislation against cloning humans, yet only a few countries passed legislation against cloning animals.

Possible Abnormalities Due to Cloning

Researchers have found several abnormalities in cloned organisms, particularly in mice. The cloned organism may be born normal and resemble its non-cloned counterpart, but majority of the time will express changes in its genome later on in life 51. The concern with cloning humans is that the changes in genomes may not only result in changes in appearance, but in psychological and personality changes as well. The theory behind this is that the biological blueprint of the genes is the same in cloned animals as it is in normal ones, but they are read and expressed incorrectly 51. DNA arrays were used to prove this claim in the research lab of Professor Rudolf Jaenisch. Jaenisch studied placentas from cloned mice and found that one in every 25 genes was expressed abnormally 52. Results of these abnormally expressed genes in the cloned mice were premature death, pneumonia, liver failure and obesity 52.

Cloning Extinct and Endangered Species

Cloning, or more precisely, the reconstruction of functional DNA from extinct species has, for decades, been a dream of some scientists. The possible implications of this were dramatized in the best-selling novel by Michael Crichton and high budget Hollywood thriller *Jurassic Park*. In real life, one of the most anticipated targets for cloning was once the Woolly Mammoth, but attempts to extract DNA from frozen mammoths have been unsuccessful, though a joint Russo-Japanese team is currently working toward this goal. And in January 2011, it was reported by Yomiuri Shimbun that a team of scientists headed by Akira Iritani of Kyoto University had built upon research by Dr. Wakayama, saying that they will extract DNA from a mammoth carcass that had been preserved in a Russian laboratory and insert it into the egg cells of an African elephant in hopes of producing a mammoth embryo. The researchers said they hoped to produce a baby mammoth within six years. In 2001, a cow named Bessie gave birth to a cloned Asian gaur, an endangered species, but the calf died after two days. In 2003, a banteng was successfully cloned, followed by three African wildcats from a thawed frozen embryo. These successes provided hope that similar techniques (using surrogate mothers of another species) might be used to clone extinct species. Anticipating this possibility, tissue samples from the last *bucardo* (Pyrenean Ibex) were frozen in liquid nitrogen immediately after it died in 2000. Researchers are also considering cloning endangered species such as the giant panda, ocelot, and cheetah. The "Frozen Zoo" at the San Diego Zoo now stores frozen tissue from the world's rarest and most endangered species.

In 2002, geneticists at the Australian Museum announced that they had replicated DNA of the Thylacine (Tasmanian Tiger), extinct about 65 years previous, using polymerase chain reaction. However, on February 15, 2005 the museum announced that it was stopping the project after tests showed the specimens' DNA had been too badly degraded by the (ethanol) preservative. On 15 May 2005 it was announced that the Thylacine project would be revived, with new participation from researchers in New South Wales and Victoria.

In January 2009, for the first time, an extinct animal, the Pyrenean ibex mentioned above was cloned, at the Centre of Food Technology and Research of Aragon, using the preserved DNA of the skin samples from 2001 and domestic goat egg-cells. (The ibex died shortly after birth due to physical defects in its lungs.) One of the continuing obstacles in the attempt to clone extinct species is the need for nearly perfect DNA. Cloning from a single specimen could not create a viable breeding population in sexually reproducing animals. Furthermore, even if males and females were to be cloned, the question would remain open whether they would be viable at all in the absence of parents that could teach or show them their natural behaviour.

Cloning endangered species is a highly ideological issue. Many conservation biologists and environmentalists vehemently oppose cloning endangered species—mainly because they think it may deter donations to help preserve natural habitat and wild animal populations. The "rule-of-thumb" in animal conservation is that, if it is still feasible to conserve habitat and viable wild populations, breeding in captivity should not be undertaken in isolation. In a 2006 review, David Ehrenfeld concluded that cloning in animal conservation is an experimental technology that, at its state in 2006, could not be expected to work except by pure chance and utterly failed a cost-benefit analysis. Furthermore, he said, it is likely to siphon funds from established and working projects and does not address any of the issues underlying animal extinction (such as habitat destruction, hunting or other overexploitation, and an impoverished gene pool). While cloning technologies are well-established and used on a regular basis in plant conservation, care must be taken to ensure genetic diversity. He concluded: Vertebrate cloning poses little risk to the environment, but it can consume scarce conservation resources, and its chances of success in preserving species seem poor. To date, the conservation benefits of transgenics and vertebrate cloning remain entirely theoretical, but many of the risks are known and documented. Conservation biologists should devote their research and energies to the established methods of conservation, none of which require transgenics or vertebrate cloning.

Blastocyst

A blastocyst is formed in the early stage of the development of an embryo. During cloning process, the blastocyst cells are obtained by scientists 5 days after the egg has been divided.

In Science Fiction

Cloning has been used in countless science fiction works throughout the years. Human cloning is usually the most popular kind that is used in these particular works, mainly due to fact that it brings up controversial questions of identity. For example, in C.J. Cherryh's novel *Cyteen*, Aldous Huxley's novel *Brave New World*, human cloning is a major plot device that not only drives the story along but also makes the reader think critically about what identity means.

Star Wars also portrays human cloning in *Star Wars Episode II: Attack of the Clones* and *Star Wars Episode III: Revenge of the Sith*, in the form of the Grand Army of the Republic, an army of clone troopers. The Expanded Universe also has numerous examples of cloning, including the Thrawn trilogy, The Hand of Thrawn duology, and Clone Wars-era media.

3

Tissue Culture

Tissue culture is the growth of tissues and/or cells separate from the organism. This is typically facilitated via use of a liquid, semi-solid, or solid growth medium, such as broth or agar. Tissue culture commonly refers to the culture of animal cells and tissues, while the more specific term plant tissue culture is being named for the plants. In 1885 Wilhelm Roux removed a section of the medullary plate of an embryonic chicken and maintained it in a warm saline solution for several days, establishing the basic principle of tissue culture. In 1907 the zoologist Ross Granville Harrison demonstrated the growth of frog nerve cell processes in a medium of clotted lymph. In 1913, E. Steinhardt, C. Israeli, and R. A. Lambert grew vaccinia virus in fragments of guinea pig corneal tissue. In modern usage, tissue culture generally refers to the growth of cells from a tissue from a multicellular organism *in vitro*. These cells may be cells isolated from a donor organism, primary cells, or an immortalised cell line. The term tissue culture is often used interchangeably with cell culture. The literal meaning of tissue culture refers to the culturing of tissue pieces, i.e. explant culture. Tissue culture is an important tool for the study of the biology of cells from multicellular organisms. It provides an *in vitro* model of the tissue in a well defined environment which can be easily manipulated and analysed.

Cell Culture

Cell culture is the complex process by which cells are grown under controlled conditions. In practice, the term "cell culture" has come to refer to the culturing of cells derived from single cellular eukaryotes, especially animal cells. However, there are also cultures of plants, fungi and microbes, including viruses, bacteria and protists. The historical development and methods of cell culture are closely interrelated to those of tissue culture and organ culture.

Animal cell culture began when scientists removed a small amount of monkey kidney from a host organism and maintained it in a warm saline solution for several days, establishing the principle of tissue culture.

Concepts in Mammalian Cell Culture

Isolation of Cells

Cells can be isolated from tissues for *ex vivo* culture in several ways. Cells can be easily purified from blood; however, only the white cells are capable of growth in culture. Mononuclear cells can be released from soft tissues by enzymatic digestion with enzymes such as collagenase, trypsin, or pronase, which break down the extracellular matrix. Alternatively, pieces of tissue can be placed in growth media, and the cells that grow out are available for culture. This method is known as explant culture. Cells that are cultured directly from a subject are known as primary cells. With the exception of some derived from tumours, most primary cell cultures have limited lifespan. After a certain number of population doublings (called the Hayflick limit), cells undergo the process of senescence and stop dividing, while generally retaining viability.

An established or immortalized cell line has acquired the ability to proliferate indefinitely either through random mutation or deliberate modification, such as artificial expression of the telomerase gene. Numerous cell lines are well established as representative of particular cell types.

Maintaining Cells in Culture

Cells are grown and maintained at an appropriate temperature and gas mixture (typically, 37°C, 5% CO_2 for mammalian cells) in a cell incubator. Culture conditions vary widely for each cell type, and variation of conditions for a particular cell type can result in different phenotypes being expressed.

Aside from temperature and gas mixture, the most commonly varied factor in culture systems is the growth medium. Recipes for growth media can vary in pH, glucose concentration, growth factors, and the presence of other nutrients. The growth factors used to supplement media are often derived from animal blood, such as calf serum. One complication of these blood-derived ingredients is the potential for contamination of the culture with viruses or prions, particularly in medical biotechnology applications. Current practice is to minimize or eliminate the use of these ingredients wherever possible and use chemically defined media, but this cannot always be accomplished. Alternative strategies involve sourcing the animal blood from countries with minimum BSE/TSE risk, such as Australia and New Zealand, and using purified nutrient concentrates derived from serum in place of whole animal serum for cell culture.

Plating density (number of cells per volume of culture medium) plays a critical role for some cell types. For example, a lower plating density makes granulosa cells exhibit estrogen production, while a higher plating density makes them appear as progesterone-producing theca lutein cells.

Cells can be grown either in suspension or adherent cultures. Some cells naturally live in suspension, without being attached to a surface, such as cells that exist in the bloodstream. There are also cell lines that have been modified to be able to survive in suspension cultures so they can be grown to a higher density than adherent conditions would allow. Adherent cells require a surface, such as tissue culture plastic or microcarrier, which may be coated with extracellular matrix components to increase adhesion properties and provide other signals needed for growth and differentiation. Most cells derived from solid tissues are adherent. Another type of adherent culture is organotypic culture, which involves growing cells in a three-dimensional (3-D) environment as opposed to two-dimensional culture dishes. This 3D culture system is biochemically and physiologically more similar to *in vivo* tissue, but is technically challenging to maintain because of many factors (e.g. diffusion).

Cell Line Cross-contamination

Cell line cross-contamination can be a problem for scientists working with cultured cells. Studies suggest anywhere from 15–20% of the time, cells used in experiments have been misidentified or contaminated with another cell line. Problems with cell line cross-contamination have even been detected in lines from the NCI-60 panel, which are used routinely for drug-screening studies. Major cell line repositories, including the American Type Culture Collection (ATCC) and the German Collection of Microorganisms and Cell Cultures (DSMZ), have received cell line submissions from researchers that were misidentified by them. Such contamination poses a problem for the quality of research produced using cell culture lines, and the major repositories are now authenticating all cell line submissions. ATCC uses short tandem repeat (STR) DNA fingerprinting to authenticate its cell lines. To address this problem of cell line cross-contamination, researchers are encouraged to authenticate their cell lines at an early passage to establish the identity of the cell line. Authentication should be repeated before freezing cell line stocks, every two months during active culturing and before any publication of research data generated using the cell lines. Many methods are used to identify cell lines, including isoenzyme analysis, human lymphocyte antigen (HLA) typing, chromosomal analysis, karyotyping, morphology and STR analysis.

One significant cell-line cross contaminant is the immortal HeLa cell line.

Manipulation of Cultured Cells

As cells generally continue to divide in culture, they generally grow to fill the available area or volume. This can generate several issues:

- Nutrient depletion in the growth media
- Accumulation of apoptotic/necrotic (dead) cells
- Cell-to-cell contact can stimulate cell cycle arrest, causing cells to stop dividing, known as contact inhibition.
- Cell-to-cell contact can stimulate cellular differentiation.

Among the common manipulations carried out on culture cells are media changes, passaging cells, and transfecting cells. These are generally performed using tissue culture methods that rely on sterile technique. Sterile technique aims to avoid contamination with bacteria, yeast, or other cell lines. Manipulations are typically carried out in a biosafety hood or laminar flow cabinet to exclude contaminating micro-organisms. Antibiotics (e.g. penicillin and streptomycin) and antifungals (e.g.amphotericin B) can also be added to the growth media.

As cells undergo metabolic processes, acid is produced and the pH decreases. Often, a pH indicator is added to the medium to measure nutrient depletion.

Media Changes

In the case of adherent cultures, the media can be removed directly by aspiration, and then is replaced.

Passaging Cells

Passaging (also known as subculture or splitting cells) involves transferring a small number of cells into a new vessel. Cells can be cultured for a longer time if they are split regularly, as it avoids the senescence associated with prolonged high cell density. Suspension cultures are easily passaged with a small amount of culture containing a few cells diluted in a larger volume of fresh media. For adherent cultures, cells first need to be detached; this is commonly done with a mixture of trypsin-EDTA; however, other enzyme mixes are now available for this purpose. A small number of detached cells can then be used to seed a new culture.

Transfection and Transduction

Another common method for manipulating cells involves the introduction of foreign DNA by transfection. This is often performed to cause cells to express a protein of interest. More recently, the transfection of RNAi constructs have been realized as a convenient mechanism for suppressing the expression of a particular gene/protein. DNA can also be inserted into cells using viruses, in methods referred to as transduction, infection or transformation. Viruses, as parasitic agents, are well suited to introducing DNA into cells, as this is a part of their normal course of reproduction.

Established Human Cell Lines

Cell lines that originate with humans have been somewhat controversial in bioethics, as they may outlive their parent organism and later be used in the discovery of lucrative medical treatments. In the pioneering decision in this area, the Supreme Court of California held in *Moore vs Regents of the University of California* that human patients have no property rights in cell lines derived from organs removed with their consent.

Generation of Hybridomas

It is possible to fuse normal cells with an immortalized cell line. This method is used to produce monoclonal antibodies. In brief, lymphocytes isolated from the spleen (or possibly blood) of an immunised animal are combined with an immortal myeloma cell line (B cell lineage) to produce a hybridoma which has the antibody specificity of the primary lymphoctye and the immortality of the myeloma. Selective growth medium (HA or HAT) is used to select against unfused myeloma cells; primary lymphoctyes die quickly in culture and only the fused cells survive. These are screened for production of the required antibody, generally in pools to start with and then after single cloning.

Applications of Cell Culture

Biological products produced by recombinant DNA (rDNA) technology in animal cell cultures include enzymes, synthetic hormones, immunobiologicals (monoclonal antibodies, interleukins, lymphokines), and anticancer agents. Although many simpler proteins can be produced using rDNA in bacterial cultures, more complex proteins that are glycosylated (carbohydrate-modified) currently must be made in animal cells. An important example of such a complex protein is the hormone erythropoietin. The cost of growing mammalian cell cultures is high, so research is underway to produce such complex proteins in insect cells or in higher plants, use of single embryonic cell and somatic embryos as a source for direct gene transfer via particle bombardment, transit gene expression and confocal microscopy observation is one of its applications. It also offers to confirm single cell origin of somatic embryos and the asymmetry of the first cell division, which starts the process.

Cell Culture in two Dimensions

Research in tissue engineering, stem cells and molecular biology primarily involves cultures of cells on flat plastic dishes. This technique is known as two-dimensional (2D) cell culture, and was first developed by Wilhelm Roux who, in 1885, removed a portion of the medullary plate of an embryonic chicken and maintained it in warm saline for several days on a flat glass plate. With the advance of polymer technology, today's standard plastic dish for 2D cell culture, commonly known as the Petri dish, was invented by Julius Richard Petri, a German bacteriologist, who is generally credited with inventing the Petri dish while working as assistant to Robert Koch.

Tissue Culture and Engineering

Cell culture is a fundamental component of tissue culture and tissue engineering, as it establishes the basics of growing and maintaining cells *ex vivo*. The major application of human cell culture is in stem cell industry, where mesenchymal stem cells can be cultured and cryopreserved for future use.

Vaccines

Vaccines for polio, measles, mumps, rubella, and chickenpox are currently made in cell cultures. Due to the H5N1 pandemic threat, research into using cell culture for influenza vaccines is being funded by the United States government. Novel ideas in the field include recombinant DNA-based vaccines, such as one made using human adenovirus (a common cold virus) as a vector, and novel adjuvants.

Culture of Non-mammalian Cells

Plant Cell Culture Methods

Plant cell cultures are typically grown as cell suspension cultures in a liquid medium or as callus cultures on a solid medium. The culturing of undifferentiated plant cells and calli requires the proper balance of the plant growth hormones auxin and cytokinin.

Bacterial and Yeast Culture Methods

For bacteria and yeasts, small quantities of cells are usually grown on a solid support that contains nutrients embedded in it, usually a gel such as agar, while large-scale cultures are grown with the cells suspended in a nutrient broth.

Viral Culture Methods

The culture of viruses requires the culture of cells of mammalian, plant, fungal or bacterial origin as hosts for the growth and replication of the virus. Whole wild type viruses, recombinant viruses or viral products may be generated in cell types other than their natural hosts under the right conditions. Depending on the species of the virus, infection and viral replication may result in host cell lysis and formation of a viral plaque.

Common Cell Lines

Human cell lines

- HeLa
- National Cancer Institute's 60 cancer cell lines
- ESTDAB database
- DU145 (prostate cancer)
- Lncap (prostate cancer)
- MCF-7 (breast cancer)
- MDA-MB-438 (breast cancer)
- PC3 (prostate cancer)
- T47D (breast cancer)
- THP-1 (acute myeloid leukemia)
- U87 (glioblastoma)
- SHSY5Y Human neuroblastoma cells, cloned from a myeloma
- Saos-2 cells (bone cancer).

Primate cell lines:

- Vero (African green monkey *Chlorocebus* kidney epithelial cell line initiated in 1962)

Rat tumor cell lines:

- GH3 (pituitary tumor)
- PC12 (pheochromocytoma)

Mouse cell lines:

- MC3T3 (embryonic calvarium)

Plant cell lines:

- Tobacco BY-2 cells (kept as cell suspension culture, they are model system of plant cell)

Other species cell lines:

- Zebrafish ZF4 and AB9 cells
- Madin-Darby canine kidney (MDCK) epithelial cell line
- Xenopus A6 kidney epithelial cells.

Biological Immortality

Biological immortality refers to a stable rate of mortality as a function of chronological age. Some individual cells and entire organisms in some species achieve this state either throughout their existence or after living long enough. This requires that death occur from injury or disease rather than deterioration, i.e., the absence of cellular senescence. However, this definition of immortality has been challenged in the new *Handbook of the Biology of Aging*, because the increase in rate of mortality as a function of chronological age may be negligible at extremely old ages, also referred to as the *late-life mortality plateau.*

The rate of mortality may cease to increase in old age, but in most cases that rate is typically very high. As a hypothetical example, there is only a 50% chance of a human surviving another year at age 110 or greater. The fact that some multicellular organisms do not experience senescence implies that aging is not an Aristotelian inevitability and refutes the prior hypothesis that aging follows a Gompertz model. No actual organism or individual cell is inviolably immortal (i.e. "invincible" or "indestructible"). Biologically immortal living things die, for example, upon receiving sufficient injury or otherwise having its body destroyed or diseased.

Cell Lines

Biologists have chosen the word immortal to designate cells that are not limited by the Hayflick limit (where cells no longer divide because of DNA damage or shortened telomeres). Prior to Leonard Hayflick, Alexis Carrel hypothesized that all normal somatic cells are immortal. The term immortalization was first applied to cancer cells that expressed the telomere-lengthening enzyme telomerase, and thereby avoided apoptosis (cell death caused by intracellular mechanisms). Among the most commonly used cell lines are HeLa and Jurkat, both of which are immortalized cancer cell lines. Normal stem cells and germ cells can also be said to be immortal (when humans refer to the cell line). Immortal cell lines of cancer cells can be created by induction of oncogenes or loss of tumor suppressor genes. One way to induce immortality is through viral-mediated induction of the large T-antigen, commonly introduced through simian virus 40 (SV-40).

Organisms

Immortality may not be desirable for multicellular organisms, as the main controls over cancer are the apoptotic mechanisms.

Tardigrades

Tardigrades, otherwise known as "water bears", are highly resilient microscopic animals. Capable of surviving extremes such as heat, radiation, drought, and even the vacuum of space by going into suspended animation, where their metabolism slows to near zero and they simply wait out the harsh conditions until the environment is more favourable. The Russian Space Administration is planning a mission to launch water bears at Phobos to study the scientific theory of panspermia.

Bacteria

Bacteria are said to be biologically immortal, but only at the level of the colony. An individual bacterium can easily die. The two daughter bacteria resulting from cell division of a parent bacterium can be regarded as unique individuals or as members of a biologically "immortal" colony. The two daughter cells can be regarded as "rejuvenated" copies of the parent cell because damaged macromolecules have been split between the two cells and diluted. In the same way stem cells and gametes can be regarded as "immortal".

Hydra

Hydras are a genus of simple, fresh-water animals possessing radial symmetry and no post-mitotic cells. The fact that all cells continually divide allows defects and toxins to be "diluted". It has been suggested that hydras do not undergo senescence (aging), and as such are biologically immortal. However, this does not explain how hydras are consequently able to maintain telomere lengths.

Jellyfish

Turritopsis nutricula is a small (5 millimeters (0.20 in)) species of jellyfish which uses transdifferentiation to replenish cells after sexual reproduction. This cycle can repeat indefinitely, potentially rendering it biologically immortal. It originated from the Caribbean sea, but has now spread around the world.

Lobsters

Older lobsters are more fertile than younger lobsters. Some scientists have claimed that they could effectively live indefinitely, barring injury, disease, capture, etc.

Life Extensionists

Some life extensionists, such as those who practice cryonics, have the hope that humans may someday become biologically immortal.

Biogerontologist Marios Kyriazis suggested that biological immortality in humans is an inevitable consequence of natural evolution. His ELPIs (Extreme Lifespans through Perpetual-equalising Interventions) theory proposes that the ability to attain indefinite lifespans is inherent in human biology, and that there will come a time when humans will continue to develop their intelligence and sophistication by living indefinitely, rather than through Darwinian evolution.

Cell Culture Assays

In biomaterials testing, a cell culture assay is any method which is used to assess the cytotoxicity of a material. This refers to the *in vitro* assessment of material to determine whether it releases toxic chemicals in sufficient quantities to kill cells either directly or indirectly through the inhibition of cell metabolic pathways. Cell culture evaluations are the precursor to whole animal studies and are a way to determine if significant cytotoxicity exists for the given material. Cell culture assays are standardised by ASTM, ISO, and BSI (British Standards Institution.)

Methods

Direct Contact Method

1. A near confluent layer of fibroblasts are prepared in a culture plate
2. Old cell culture media (agar generally) is removed
3. Fresh media is added
4. Material being tested is placed onto the cultures, which are incubated for 24 hours at 37 degrees Celsius
5. The material is removed
6. The culture media is removed
7. The remaining cells are fixed and stained, dead cells are lost during fixation and only the live cells are stained
8. The toxicity of the material is indicated by the absence of stained cells around the material.

Agar Diffusion Method

1. A near confluent layer of fibroblasts are prepared in a culture plate
2. Old cell culture media is removed
3. The cells are covered with a solution of 2% agar, which often contains red vital stain
4. When the agar solidifies the cells will have dispersed throughout its volume
5. The material is then placed on the surface of the agar and incubated for 24 hours at 37 degrees Celsius
6. Live cells take up the vital stain and retain it, dead cells do not
7. The toxicity of the material is evaluated by the loss of vital stain under and around the material
8. Surface microscopy is also needed to evaluate the material-cell interface.

Elution Method

1. A near confluent layer of fibroblasts are prepared in a culture plate
2. An extract of the material which is being tested is prepared using physiological saline or serum free media (the latter is generally preferred)

3. Extraction conditions are used which are appropriate for the type of exposure which the cells would receive in the *in vivo* environment if the material were to be implanted
4. The extract is placed on the cells and incubated for 48 hours at 37 degrees Celsius
5. After 48 hours the toxicity is evaluated using either a histochemical or vital stain.

Each method has its own advantages and disadvantages, and some are more suitable for certain applications than others. For example the direct contact method offers conditions which are most similar to the physiological environment but the cells are susceptible to trauma if the material moves. The agar diffusion method is good for materials with high densities and offers an even concentration gradient for potential toxicants, but there is a serious risk of the cells going into thermal shock when they are overlayed with agar. The elution method is best for applications which might require extra incubation time, but additional time and steps are required for preparing such a test.

In vitro biomaterials testing yields fundamental information about the behaviour of materials in contact with living cells, but cannot qualify or even accurately predict the performance of a material *in vivo*.

Electric Cell-substrate Impedance Sensing

Electric cell-substrate impedance sensing or ECIS (a trademark of Applied BioPhysics Inc.) refers to a non-invasive biophysical approach to monitor living animal cells *in vitro*, i.e. within a well-defined laboratory environment.

In ECIS the cells are grown on the surface of small and planar gold-film electrodes, which are deposited on the bottom of a cell culture dish (petri dish). The AC impedance of the cell-covered electrode is then measured at one or several frequencies as a function of time. Due to the insulating properties of their membranes the cells behave like dielectric particles so that the impedance increases with increasing coverage of the electrode until a confluent (i.e. continuous) layer of cells is established. In confluent cell layers the measured impedance is mainly determined by the three-dimensional shape of the cells. If cell shape changes occur, the current pathways through and around the cell bodies change as well, leading to a corresponding increase or decrease of impedance. Thus, by recording time-resolved impedance measurements, cell shape changes can be followed in real time with sub-microscopic resolution and can be used for bioanalytic purposes.

As the shape of animal cells responds very sensitively to alterations in metabolism as well as chemical, biological or physical stimuli, the ECIS technique is applied in various different experimental settings in cell biological research laboratories. It can be used as a sensor in cytotoxicity studies, drug development or as a non-invasive means to follow cell adhesion to *in vitro* surfaces. Equipments based on the ECIS technique are also dedicated to monitor the chemokinetic activity of adherent cells spread on the electrode surface (micromotion) as well as their chemotactic activities in ECIS-based wound healing assays.

Plant Tissue Culture

Plant tissue culture is a collection of techniques used to maintain or grow plant cells, tissues or organs under sterile conditions on a nutrient culture medium of known composition. Plant tissue culture is widely used to produce clones of a plant in a method known as micropropagation. Different techniques in plant tissue culture may offer certain advantages over traditional methods of propagation, including:

- The production of exact copies of plants that produce particularly good flowers, fruits, or have other desirable traits.
- To quickly produce mature plants.
- The production of multiples of plants in the absence of seeds or necessary pollinators to produce seeds.
- The regeneration of whole plants from plant cells that have been genetically modified.
- The production of plants in sterile containers that allows them to be moved with greatly reduced chances of transmitting diseases, pests, and pathogens.
- The production of plants from seeds that otherwise have very low chances of germinating and growing, i.e.: orchids and nepenthes.
- To clean particular plants of viral and other infections and to quickly multiply these plants as 'cleaned stock' for horticulture and agriculture.

Plant tissue culture relies on the fact that many plant cells have the ability to regenerate a whole plant (totipotency). Single cells, plant cells without cell walls (protoplasts), pieces of leaves, or (less commonly) roots can often be used to generate a new plant on culture media given the required nutrients and plant hormones.

Techniques

Modern plant tissue culture is performed under aseptic conditions under filtered air. Living plant materials from the environment are naturally contaminated on their surfaces (and sometimes interiors) with microorganisms, so surface sterilization of starting materials (explants) in chemical solutions (usually Sodium or calcium hypochlorite or mercuric chloride) is required. Mercuric chloride is seldom used as a plant sterilant today, unless other sterilizing agents are found to be ineffective, as it is dangerous to use, and is difficult to dispose of. Explants are then usually placed on the surface of a solid culture medium, but are sometimes placed directly into a liquid medium, particularly when cell suspension cultures are desired. Solid and liquid media are generally composed of inorganic salts plus a few organic nutrients, vitamins and plant hormones. Solid media are prepared from liquid media with the addition of a gelling agent, usually purified agar.

In-vitro Tissue Culture Potato Explants

The composition of the medium, particularly the plant hormones and the nitrogen source (nitrate versus ammonium salts or amino acids) have profound effects on the morphology of the tissues that grow from the initial explant. For example, an excess of auxin will often result in a proliferation of roots, while an excess of cytokinin may yield shoots. A balance of both auxin and cytokinin will often produce an unorganised growth of cells, or callus, but the morphology of the outgrowth will depend on the plant species as well as the medium composition.

As cultures grow, pieces are typically sliced off and transferred to new media (subcultured) to allow for growth or to alter the morphology of the culture. The skill and experience of the tissue culturist are important in judging which pieces to culture and which to discard. As shoots emerge from a culture, they may be sliced off and rooted with auxin to produce plantlets which, when mature, can be transferred to potting soil for further growth in the greenhouse as normal plants.

Choice of Explant

The tissue obtained from the plant to culture is called an explant. Based on work with certain model systems, particularly tobacco, it has often been claimed that a totipotent explant can be grown from any part of the plant. However, this concept has been vitiated in practice. In many species explants of various organs vary in their rates of growth and regeneration, while some do not grow at all. The choice of explant material also determines if the plantlets developed via tissue culture are haploid or diploid. Also the risk of microbial contamination is increased with inappropriate explants. Thus it is very important that an appropriate choice of explant be made prior to tissue culture.

The specific differences in the regeneration potential of different organs and explants have various explanations. The significant factors include differences in the stage of the cells in the cell cycle, the availability of or ability to transport endogenous growth regulators, and the metabolic capabilities of the cells. The most commonly used tissue explants are the meristematic ends of the plants like the stem tip, auxiliary bud tip and root tip. These tissues have high rates of cell division and either concentrate or produce required growth regulating substances including auxins and cytokinins.

The pathways through which whole plants are regenerated from cells and tissues or explants such as meristems broadly fall into three types:

1. The method in which explants that include a meristem (viz. the shoot tips or nodes) are grown on appropriate media supplemented with plant growth regulators to induce proliferation of multiple shoots, followed by rooting of the excised shoots to regenerate whole plants,
2. The method in which totipotency of cells is realised in the form of *de novo* organogenesis, either directly in the form of induction of shoot meristems on the explants or indirectly via a callus (unorganised mass of cells resulting from proliferation of cells of the explant) and plants are regenerated through induction of roots on the resultant shoots,
3. Somatic embryogenesis, in which asexual adventive embryos(comparable to zygotic embryos in their structure and development) are induced directly on explants or indirectly through a callus phase.

The first method involving the meristems and induction of multiple shoots is the preferred method for the micropropagation industry since the risks of somaclonal variation (genetic variation induced in tissue culture) are minimal when compared to the other two methods. Somatic embryogenesis is a method that has the potential to be several times higher in multiplication rates and is amenable to handling in liquid culture systems like bioreactors. Some explants, like the root tip, are hard to isolate and are contaminated with soil microflora that become problematic during the tissue culture process. Certain soil microflora can form tight associations with the root systems, or even grow within the root. Soil particles bound to roots are difficult to remove without injury to the roots that then allows microbial attack. These associated microflora will generally overgrow the tissue culture medium before there is significant growth of plant tissue.

Aerial (above soil) explants are also rich in undesirable microflora. However, they are more easily removed from the explant by gentle rinsing, and the remainder usually can be killed by surface sterilization. Most of the surface microflora do not form tight associations with the plant tissue. Such associations can usually be found by visual inspection as a mosaic, decolourization or localized necrosis on the surface of the explant. An alternative for obtaining uncontaminated explants is to take explants from seedlings which are aseptically grown from surface-sterilized seeds. The hard surface of the seed is less permeable to penetration of harsh surface sterilizing agents, such as hypochlorite, so the acceptable conditions of sterilization used for seeds can be much more stringent than for vegetative tissues.

Tissue cultured plants are clones, if the original mother plant used to produce the first explants is susceptible to a pathogen or environmental condition, the entire crop would be susceptible to the same problem, conversely any positive traits would remain within the line also.

Applications

Plant tissue culture is used widely in plant science; it also has a number of commercial applications. Applications include:

- Micropropagation is widely used in forestry and in floriculture. Micropropagation can also be used to conserve rare or endangered plant species.
- A plant breeder may use tissue culture to screen cells rather than plants for advantageous characters, e.g. herbicide resistance/tolerance.
- Large-scale growth of plant cells in liquid culture in bioreactors for production of valuable compounds, like plant-derived secondary metabolites and recombinant proteins used as biopharmaceuticals.
- To cross distantly related species by protoplast fusion and regeneration of the novel hybrid.
- To cross-pollinate distantly related species and then tissue culture the resulting embryo which would otherwise normally die (Embryo Rescue).
- For production of doubled monoploid (dihaploid) plants from haploid cultures to achieve homozygous lines more rapidly in breeding programmes, usually by treatment with colchicine which causes doubling of the chromosome number.
- As a tissue for transformation, followed by either short-term testing of genetic constructs or regeneration of transgenic plants.
- Certain techniques such as meristem tip culture can be used to produce clean plant material from virused stock, such as potatoes and many species of soft fruit.
- micropropagation using meristem and shoot culture to produce large numbers of identical individuals.
- production of identical sterile hybrid species can be obtained.

Laboratories

Although some growers and nurseries have their own labs for propagating plants by the technique of tissue culture, a number of independent laboratories provide custom propagation services. The Plant Tissue Culture Information Exchange lists many commercial tissue culture labs. Since plant tissue culture is a very labour intensive process, this would be an important factor in determining which plants would be commercially viable to propagate in a laboratory.

Organ Culture

Organ culture is a development from tissue culture methods of research, the organ culture is able to accurately model functions of an organ in various states and conditions by the use of the actual *in vitro* organ itself. Parts of an organ or a whole organ can be cultured in vitro. The main objective is to maintain the architecture of the tissue and direct it towards normal development. In this technique, it is essential that the tissue is never be disrupted or damaged. It thus requires careful handling. The media used for a growing organ culture are generally the same as those used for tissue culture. The techniques for organ culture can be classified into (i) those employing a solid medium and (ii) those employing liquid medium.

Methodology

Embryonic organ culture is an easier alternative to normal organ culture derived from adult animals. The following are three techniques employed for embryonic organ culture.

Plasma Clot Method

The following are general steps in organ culture on plasma clots.

1. Prepare a plasma clot by mixing 15 drops of plasma with five drops of embryo extract in a watch glass.
2. Place a watch glass on a pad of cotton wool in a petri dish; cotton wool is kept moist to prevent excessive evaporation from the dish.
3. Place a small, carefully dissected piece of tissue on top of the plasma clots in watch glass.

The technique has now been modified, and a raft of lens paper or rayon net is used on which the tissue is placed. Transfer of the tissue can then be achieved by raft easily. Excessive fluid is removed and the net with the tissue placed again on the fresh pool of medium. Parts of an organ or a whole organ can be cultured in vitro.

The main objective is to maintain the architecture of the tissue and direct it towards normal development. In this technique, it is essential that the tissue never be disrupted or damaged. It thus requires careful handling.

The media used for a growing organ are generally the same as those used for tissue culture. The techniques for organ culture can be classified into (i) those employing a solid medium and (ii) those employing liquid medium.

Agar Gel Method

Media solidified with agar are also used for organ culture and these media consist of 7 parts 1% agar in BSS, 3 parts chick embryo extract and 3 parts of horse serum. Defined media with or without serum are also used with agar.

The medium with agar provides the mechanical support for organ culture. It does not liquefy. Embryonic organs generally grow well on agar, but adult organ culture will not survive on this medium. The culture of adult organs or parts from adult animals is more difficult due to their greater requirement of oxygen. A variety of adult organs (e.g. the liver) have been cultured using special media with special apparatus (Towell's II culture chamber). Since serum was found to be toxic, serum-free media were used, and the special apparatus permitted the use of 95% oxygen.

Raft Methods

In this approach the explant is placed onto a raft of lens paper or rayon acetate, which is floated on serum in a watch glass. Rayon acetate rafts are made to float on the serum by treating their 4 corners with silicone. Similarly, floatability of lens paper is enhanced by treating it with silicone. On each raft, 4 or more explants are usually placed. In a combination of raft and clot techniques, the explants are first placed on a suitable raft, which is then kept on a plasma clot. This modification makes media changes easy, and prevents the sinking of explants into liquefied plasma.

Grid Method

Initially devised by Trowell in 1954, the grid method utilises 25 mm x 25 mm pieces of a suitable wire mesh or perforated stainless steel sheet whose edges are bent to form 4 legs of about 4 mm height. Skeletal tissues are generally placed directly on the grid but softer tissues like glands or skin are first placed on rafts, which are then kept on the grids.

The grids themselves are placed in a culture chamber filled with fluid medium up to the grid; the chamber is supplied with a mixture of O_2 and CO_2 to meet the high O_2 requirements of adult mammalian organs. A modification of the original grid method is widely used to study the growth and differentiation of adult and embryonic tissues.

Limitations

- Results from organ cultures are often not comparable to those from whole animals studies, e.g. in studies on drug action since the drugs are metabolised in vivo but not in vitro.

Current Progress

In April 2006, scientists reported a successful trial of seven bladders grown in-vitro and given to humans. A bladder has been cultured by Anthony Atala of the Wake Forest Institute for Regenerative Medicine in Winston-Salem, North Carolina. A jawbone has been cultured at Columbia University, a lung has been cultured at Yale. A beating rat heart has been cultured by Doris Taylor at the University of Minnesota. An artificial kidney has been cultured by H. David Humes at the University of Michigan.

Animals that have been Cloned

This is a list of animals that have been cloned in alphabetical order. One significant aspect of this list is documenting the transition from early concerns that animal cloning procedures might be limited to a few species, that cloned animals might be physiologically abnormal, or cloning might lack utility for societ.

Carp

Chinese embryologist Tong Dizhou successfully inserted the DNA from a male Asian carp into the egg of a female Asian carp to create the first fish clone in 1963. In 1973 Dizhou inserted Asian carp DNA into a European crucian carp to create the first interspecies clone.

Cat

- In December 2001, scientists at Texas A&M University created the first cloned cat, CC (CopyCat). Even though CC is an exact copy of his host, they have different personalities; i.e. CC is shy and timid, his host on the other hand is playful and curious.
- In 2004, the first commercially cloned cat, Little Nicky, was created by Genetic Savings & Clone.

Cattle

- First World cloned calf (Gene) was born on February 7, 1997 on American Breeders Service facilities in Deforest, Wisconsin. Later it was transferred and kept to Minnesota Zoo Education Centre.
- A Holstein heifer named Amy was cloned by Dr. Xiangzhong (Jerry) Yang using ear skin cells from a high-merit cow named Aspen at the University of Connecticut on June 10, 1999, followed by three additional clones, Betty, Cathy and Daisy by July 7, 1999.
- Second Chance, a Brahman bull was cloned from Chance, a beloved celebrity bull. Second Chance was born August 9, 1999 at Texas A&M University.
- Texas A&M University cloned a Black Angus bull named 86 Squared in 2000, after cells from his donor, Bull 86, had been frozen for 15 years. Both bulls exhibit a natural resistance to Brucellosis, Tuberculosis and other diseases which can be transferred in meat.
- Millie and Emma were two female Jersey cows cloned at the University of Tennessee in 2001. They were the first cows to be produced using standard cell-culturing techniques.
- Pampa the first animal cloned in Argentina by Biosidus (2002)
- Ten more Jersey cows were cloned at the University of Tennessee. (females, 2002)

- Bonyana and Tamina cloned calf in Royan Research Institute, Isfahan, Iran in summer of 2009.
- In 2010 the first Spanish Fighting Bull was cloned by Spanish scientists.
- Anatolian Grey bull (Efe) was cloned in Turkey in 2009 and cattle from the same breed (Ece, Ecem, Nilufer, Kiraz) by TUBITAK.
- GARIMA- I: world's first buffalo calf through the "Hand guided Cloning Technique" was born on February 6, 2009 at NDRI, Karnal(India).
- GARIMA- II: NDRI, Karnal(India).
- Cloned male buffalo calf Shresth born on August 26, 2010 at National Dairy Research Institute, Karnal, India-132001.

Deer

- Dewey was born on May 23, 2003 at Texas A&M University.

Dog

- South Korean scientist Hwang Woo-Suk cloned the first dog, an afghan hound named Snuppy. Later in 2005 Hwang Woo-Suk was found to have fabricated evidence in stem cell research projects. This caused some to question the veracity of his other experiments, including Snuppy. In their investigation of Hwang Woo-Suk's publication, however, a team from SNU confirmed that Snuppy was a true clone of Tei, the DNA donor dog. South Korean scientists recently cloned 'sniffer' dogs.
- BioArts International held a dog cloning contest where people would send in submissions about which dog was the most suited to be cloned. The winner was Trakr, a K-9 police dog who was a 9/11 hero.
- In summer 2011, South Korean researchers cloned a beagle dog named Tegon, which glowed in ultraviolet light

Ferret

Clones Libby and Lilly were produced via nuclear transfer by cell fusion in 2004.

Frog (Tadpole)

In 1958, John Gurdon, then at Oxford University, explained that he had successfully cloned a frog. He did this by using intact nuclei from somatic cells from a Xenopus tadpole. This was an important extension of work of Briggs and King in 1952 on transplanting nuclei from embryonic blastula cells

Fruit Flies

- (2004).

Gaur

A species of wild cattle, the first endangered species to be cloned. In 2001 at the Trans Ova Genetics in Sioux Centre, Iowa, USA, a cloned Gaur was born from a surrogate domestic cow mother. However, the calf died within 48 hours.

Goat

- Downen TX 63 684 (nicknamed Megan) was cloned from a top producing Boer goat born on March 29, 2001 at Texas A&M University.
- The Middle East's first and the world's fifth cloned goat, 'Hanna', has been successfully born at Royan institute in Isfahan, Iran. The cloned goat was developed in the surrogate uterus of a black Bakhtiari goat for 147 days and was born, Wednesday, at 1:30 a.m. through a cesarean section. She is reported to be in a good health. Hanna, also known as R-CAP-C1, is completely distinguished from other goats because of its white and henna-like colour. Iran's first cloned lamb, Royana, was born September 30, 2006 in Royan institute and was able to survive the post-natal complications common in cloned animals. Iranian researchers are looking to use cloned goats to produce the genetically modified animals required for manufacturing new recombinant medications.

Horse

- Prometea, female, born May 2003, Italy
- Pieraz, male, born February 2005, Italy
- Paris-Texas, male, born March 2005, USA
- Gemini, male, born September 2008, USA, clone of multiple recipient of "Horse of the Year" award for jumping Gem Twist
- Saphir, male, born February 2010, USA, clone of show jumper Sapphire.

Mice

- Possibly the first cloned mammals was a mouse (named "Masha") in 1986, in the Soviet Union. However, the cloning was done from an embryo cell, while the sheep Dolly in 1996 was cloned from an adult cell.
- The first mouse from adult cells, Cumulina, was born in 1997 at the University of Hawai'i at Mânoa in the laboratory of Ryuzo Yanagimachi using the Honolulu technique.
- Over a dozen clones as of 2002.

Mouflon

- An endangered species, the Mouflon was the first to live past infancy. Cloned 2001

Mule

- Idaho Gem (male, May 2003)
- Utah Pioneer (male, June 2003)
- Idaho Star (male, July 2003).

Pig

- 5 Scottish PPL piglets (Millie, Alexis, Dotcom, Carrel, and Christa) (March 5, 2000)

Pyrenean Ibex

In 2009, one clone was born alive, but died seven minutes later, due to physical defects in the lungs. The Pyrenean Ibex became the first taxon ever to come back from extinction, for a period of seven minutes in January 2009.

Rabbit

- In France (March–April, 2003).

Rat

- Ralph (male, 2003).

Rhesus Monkey

- Tetra (female, January 2000) by embryo splitting.
- Cloned embryos (November 2007) by tranfer of DNA from adult cells.

Sheep

- From early embryonic cells by Steen Willadsen (1986). Megan and Morag cloned from differentiated embryonic cells in June 1995.
- Dolly (1996–2003), first cloned mammal from somatic cells
- Polly and Molly (July 1997), first transgenic cloned mammal
- Royana (2006) cloned in Royan Research institute in Isfahan, Iran.
- Oyalý and Zarife were cloned in November 2007 in Istanbul University in Istanbul, Turkey.

Water Buffalo

The world's first water buffalo was cloned either in Beijing China in 2005 or New Delhi, India in 2009 "Samrupa", the world's first cloned buffalo calf, which died a week later from a lung infection.

Wolf

- An endangered species of wolf cloned by Korean scientists including the controversial scientist Hwang Woo-Suk
- There are two cloned wolves in a zoo in Korea for public view, they are called Snuwolf and Snuwolffy which are names taken from the university in Korea, Seoul National University.

Fish Cloning

Researchers are pondering how an exclusively-female fish species has survived without genetically self-destructing itself into extinction, the BBC reports. The Amazon Molly, found in Mexico and Texas, reproduces by cloning, and while it does "interact" with chaps from other species to "trigger its reproduction process", doesn't utilise any of their DNA in producing offspring.

While this might be considered proof of the old saw that "a woman needs a man like a fish needs a bicycle", scientists from the University of Edinburgh applied "complex mathematical models" to demonstrate that the Amazon Molly should in fact have become extinct within the last 70,000 years as a result of the adverse genetic changes resulting from asexual reproduction.

Quite how it has survived remains a mystery, although one theory suggests occasional liaisons with males in which DNA is taken on board to "refresh" the species' genetic make-up.

Dr Laurence Loewe, of the university's School of Biological Sciences, said: "What we have shown now is that this fish really has something special going on and that some special tricks exist to help this fish survive. Maybe there is still occasional sex with strangers that keeps the species alive. Future research may give us some answers."

Holly, the World's First Semi-cloned Fish

Some 8 years ago, semi-cloning was considered a scientific fiction and 26 years ago, using mice for the experiment, British scientist Sir Martin Evans attempted to make haploid embryonic stem (ES) cells. Fast-forward to 2009 and it is now a team led by Assoc Prof Hong Yunhan of the Department of Biological Sciences, who have successfully produced the world's first haploid embryonic stem cells and semi-cloned fish. Their work was published in the 16 October issue of the highly acclaimed *Science* magazine.

"While traditional cloning uses a 'replacement' strategy, semi-cloning adopts a 'combination' strategy. In semi-cloning, scientists combine two haploid nuclei - one haploid nucleus from the embryonic stem cell mimicking the sperm nucleus, and the nucleus of an unfertilized egg (from the mother) - using nuclear transfer," said Assoc Prof Hong.

> *In the practice of nuclear transfer for the production of Holly, we introduced a haploid embryonic stem cell nucleus into a mature egg. In this way, scientists mimic the reproduction process artificially, creating offspring that carry genetic traits from both parents similar to bisexual reproduction,* he added.

For the study, the team worked the Medaka fish as it shares a number of genes with humans in the area of reproduction. The Medaka fish is also easy to maintain being small in size and producing 30 to 40 eggs a day. Experimenting with fish also ensured that there would be no ethical issues.

The findings of Assoc Prof Hong have important implications for reproductive medicine and technology. With semi-cloning which is 50 to 100 times more efficient than cloning, there is hope for the treatment of infertility, in particular male infertility which is more prominent. In addition, the NUS team's study has generated haploid embryonic stem cells which could be used to study the effect of recessive mutations of essential genes that may not be apparent in normal diploid cells. This spells the possibility of developing mammalian haploid embryonic stem cell cultures for disease analysis.

Assoc Prof Hong said that the study using the fish model took some five years and a funding of SGD1.5 million to complete. This included time for consolidation of ideas, practical and administrative work. The next step would be to conduct the study on mice which will take approximately another five years. He is also currently looking for partners in Asia for collaboration. A team of scientists at the Michigan State University has found a new, more efficient method for cloning zebra fish.

"After the mouse, it is the most commonly used vertebrate in genetic studies," said Jose Cibelli, an MSU professor of animal science and one of the paper's co-authors. "It is used in cancer research and cardiovascular research because they have many of the same genes we have."

Zebra fish is also used by scientists researching normal development and birth defects, as well as various human diseases and the functions of cell populations within organs.

Up until now, zebra fish cloning has had a low success rate, but the new Michigan method has changed this. The new method uses ovarian fluid from a Chinook salmon to keep the unfertilized egg alive.

"This worked well, because it kept the egg inactive for some time", Cibelli explained. "It gave us two or three hours to work with".

During the next step of the process, the Michigan researchers used a laser to remove DNA from the egg; a method borrowed from human in vitro fertilization. Next, the team devised a new, more efficient way of inserting the donor cells into the egg.

Cytoplasmic Impact on Cross-Genus Cloned Fish Derived from Transgenic Common Carp (Cyprinus carpio) Nuclei and Goldfish (Carassius auratus) Enucleated Eggs

Most cloned animals were obtained by intraspecies nuclear transfer and are phenotypically identical to their nuclear donors; furthermore, there was no further report on successful fish cloning since the report of cloned zebrafish. Here we report the production of seven cross-genus cloned fish by transferring nuclei from transgenic common carp into enucleated eggs of goldfish. Nuclear genomes of the cloned fish were exclusively derived from the nuclear donor species, common carp, whereas the mitochondrial DNA from the donor carp gradually disappeared during the development of nuclear transfer (NT) embryos. The somite development process and somite number of nuclear transplants were consistent with the recipient species, goldfish, rather than the nuclear donor species, common carp. This resulted in a long-lasting effect on the vertebral numbers of the cloned fish, which belonged to the range of goldfish. These demonstrate that fish egg cytoplasm not only can support the development driven by transplanted nuclei from a distantly related species at the genus scale but also can modulate development of the nuclear transplants.

With the advent of cloned livestock, yet another biotech science experiment may soon find its way to the American dinner table. In January 2008, the FDA essentially told the public that the meat and milk from cloned livestock are safe for human consumption. FDA's action flies in the face of widespread scientific concern about the risks of food from clones, and ignores the animal cruelty and troubling ethical concerns that the cloning process brings. The approval also goes against the will of Congress, who voted twice in 2007 to delay FDA's decision on cloned animals until additional safety and economic studies can be completed, and ignores the feelings of the American public, 150,000 of whom wrote to FDA opposing the approval during last year's public comment period. What is worse, FDA will not require labelling on cloned food, so consumers will have no way to avoid these experimental foods.

Animal cloning is a new technology with potentially severe risks for food safety. Defects in clones are common, and cloning scientists warn that even small imbalances in clones could lead to hidden food safety problems in clones' milk or meat. There are few studies on the risks of food from clones, and no long-term food safety studies have been done. Numerous opinion polls show that the majority of Americans do not want food from animal clones and are opposed to cloning on moral or ethical grounds. The FDA's veterinary medicine advisory panel rebuked the agency in 2003 for its position, declaring that not enough research has been done to determine whether food derived from cloned animals is safe. In fact, livestock cloning raises numerous health and ethical concerns. Over 90 percent of cloning attempts fail, and cloned animals that are born have more health problems and higher mortality rates than sexually reproduced animals.

Given that researchers do not understand many of the health problems that arise throughout the life cycles of cloned animals, the FDA acted irresponsibly in assuming that the foods produced from these animals are safe for humans to eat. According to Ian Wilmut, the leader of the team of scientists that cloned the sheep Dolly, determining the health impacts of food derived from clones must be based on the animals' complete health profiles. Such studies have not been done.

Amphibian Cloning

Since the production of Dolly the sheep cloning methods for somatic cells have been thoroughly described and are becoming routine. However, the rate at which live clones are produced remains low in all mammalian species tested so far. Remarkably, irrespective of the cloning protocol or the donor-cell type, all clones display common abnormalities, particularly in the placenta. The process is also complicated by early mortality of somatic-cell clones and the founder mammalian clone, Dolly the sheep, died in February 2003 aged six years. Based on published data and on our own experience, our view is that mammalian somatic-cell cloning and the pioneer nuclear-transfer data from amphibians have much in common. We suggest that the only way to improve nuclear reprogramming is to modify the chromatin structure of somatic cells before nuclear transfer, to provide the oocyte with a chromosomal structure that is more compatible with the natural reprogramming machinery of the oocyte.

The ultimate test of whether the nucleus of a differentiated cell has undergone any irreversible functional restriction is to have that nucleus generate every other type of differentiated cell in the body. If each cell's nucleus is identical to the zygote nucleus, then each cell's nucleus should also be totipotent (capable of directing the entire development of the organism) when transplanted into an activated enucleated egg.

As early as 1895, the embryologist Yves Delage predicted that "If, without deterioration, the egg nucleus could be replaced by the nucleus of an ordinary embryonic cell, we should probably see this egg developing without changes". Before such an experiment could be done, however, three techniques for transplanting nuclei into eggs had to be perfected: (1) a method for enucleating host eggs without destroying them; (2) a method for isolating intact donor nuclei; and (3) a method for transferring such nuclei into the host egg without damaging either the nucleus or the oocyte.

All three techniques were developed in the 1950s by Robert Briggs and Thomas King. First, they combined the enucleation of the host egg with its activation. When an oocyte (a developing egg cell) from the leopard frog *Rana pipiens* is pricked with a clean glass needle, the egg undergoes all the cytological and biochemical changes associated with fertilization. The internal cytoplasmic rearrangements of fertilization occur, and the completion of meiosis takes place near the animal pole of the cell. The meiotic spindle can easily be located as it pushes away the pigment granules at the animal pole, and puncturing the oocyte at this site causes the spindle and its chromosomes to flow outside the egg. The host egg is now considered both activated (the fertilization reactions necessary to initiate development have been completed) and enucleated.

The transfer of a nucleus into the egg is accomplished by disrupting a donor cell and transferring the released nucleus into the oocyte through a micropipette. Some cytoplasm accompanies the nucleus to its new home, but the ratio of donor to recipient cytoplasm is only $1:10^5$, and the donor cytoplasm does not seem to affect the outcome of the experiments. In 1952, Briggs and King, using these techniques, demonstrated that blastula cell nuclei could direct the development of complete tadpoles when transferred into the cytoplasm of an activated enucleated frog egg. This procedure is called somatic nuclear transfer, or more commonly, cloning. What happens when nuclei from more advanced developmental stages are transferred into activated enucleated oocytes? King and Briggs (1956) found that whereas most blastula nuclei could produce entire tadpoles, there was a dramatic decrease in the ability of nuclei from later stages to direct development to the tadpole stage. When nuclei from the somatic cells of tailbud-stage tadpoles were used as donors, normal development did not occur. However, nuclei from the germ cells of tailbud-stage tadpoles (which could give rise to a complete organism after fertilization) were capable of directing normal development in 40 percent of the blastulae that developed (Smith, 1956). Thus, most somatic cells appeared to lose their ability to direct development as they became determined and differentiated.

Amphibian Cloning: The Totipotency of Somatic Cells

Is it possible that some differentiated cell nuclei differ from others in their ability to direct development? John Gurdon and his colleagues, using slightly different methods of nuclear transplantation on the frog *Xenopus*, obtained results suggesting that the nuclei of some differentiated cells can remain totipotent. Gurdon, too, found a progressive loss of potency with increasing developmental age, although *Xenopus* cells retained their potencies for a longer period than did the cells of *Rana*.

To clone amphibians from the nuclei of cells known to be differentiated, Gurdon and his colleagues cultured epithelial cells from adult frog foot webbing. These cells were shown to be differentiated: each of them contained a specific keratin, the characteristic protein of adult skin cells. When nuclei from these cells were transferred into activated enucleated *Xenopus* oocytes, none of the first-generation transfers progressed further than the formation of the neural tube shortly after gastrulation. By serial transplantation (i.e., taking nuclei from the cloned blastulas), however, numerous tadpoles were generated (Gurdon et al., 1975). Although the tadpoles all died prior to feeding, they showed that a single differentiated cell nucleus still retained incredible potencies. A nucleus of a skin cell could produce all the cells of the young tadpole.

Crmtin Diminti CMI DI CD

Chromosomal diminution (or more properly, chromatin diminution) is a fascinating exception to the general rule of the constancy of the genome. Moreover, it provides an interesting alternative to differential gene expression. Instead of repressing genes that are not being used, why not get rid of them altogether? In some species, entire chromosomes are eliminated during development (chromosome elimination). In chromatin diminution, selected portions of the chromosomes are cast away in particular cells (Müller et al., 1996).

While chromatin diminution was observed by Boveri over 100 years ago, it can now be studied on the molecular level. This has been accomplished in the roundworm *Ascaris suum* (formerly *Ascaris lumbricoides* var. *suum*), a parasite in the intestines of pigs. (Large quantities of this worm can be obtained cheaply from slaughterhouses all over the world.) These studies reveal that chromatin diminution is not a haphazard ripping apart of the genome. Rather, this diminution is a carefully orchestrated genome rearrangement, where certain regions of the DNA are degraded and new telomeres are constructed.

In *A. suum*, all the somatic cell precursors undergo chromatin diminution. The process differs slightly from the one that Boveri observed in *Parascalis univalens*. First, in *A. suum*, diminution was not seen before the third cell division. Second, there is some chromosomal fragmentation in the germline cells of *Ascaris*, and this results in the loss of some telomeric chromatin (Tobler, 1986). Third, in *Parascaris*, about 85% of the total nuclear DNA is eliminated. In *Ascaris*, this figure is closer to 25%. (Moritz and Roth, 1976). In both species, all the heterochromatin is lost from the somatic cell nuclei. Chromatin diminution is not a general property of roundworms, nor is it in all species of parasitic roundworms. In the free-living nematode *Caenorhabditis elegans*, there is no known chromatin diminution (Emmons et al., 1979).

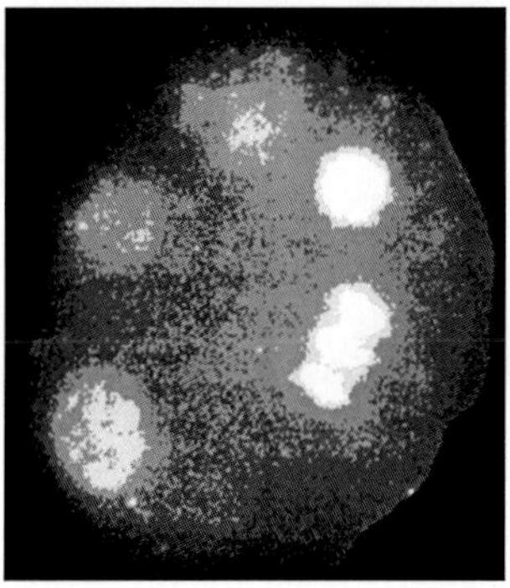

Figure : *Elimination mitosis in an early* Ascaris *embryo. The DNA is stained light blue with DAPI. One of the cells is undergoing mitosis. The eliminated chromatin is left at the equatorial plate of mitosis while the other chromatin is being drawn to the mitotic poles. (After Müller et al., 1991.)*

Chromatin diminution in *A. suum* occurs at specific chromosomal breakage regions. Nothing is known about the mechanism of this breakage or the selection of these sites. The sites have no obvious consensus sequence. The DNA between the break site and the telomere (any thus gene in this region) is degraded. After breakage, multiple units of the telomeric sequence TTAGGC are added to the break site. The microtubules of the spindle apparatus associate only with non-eliminated chromosomal regions.

Moreover, the heteromeric, eliminated chromatin lacks kinetochore plates, thereby causing them to remain at the equatorial plate and not be drawn into the mitotic poles. Here, they are attacked by degrading enzymes and eventually disintegrate.

The eliminated DNA is enriched in repetitive DNA sequences rather than in the single-copy DNA that contains most genes. However, two single-copy genes have recently been found to become eliminated from the somatic cell precursors of *Ascaris suum*. The *fert-1* gene is on soon after fertilization, and it makes several types of transcripts. Two of these transcripts are long-lived and persist in the somatic cells long after the genes that encode it have been eliminated (Spicher et al., 1994). Thus, the longevity of the message may compensate for the loss of the gene. The second gene is that encoding a germline-specific isoform of the S19 ribosomal protein. It turns out that *Ascaris* has two types of S19 genes. Both are active in the germ line, while only one is active in the somatic cells (Etter et al., 1994).

4

Chromosomes in Cells

A chromosome is an organized structure of DNA and protein found in cells. It is a single piece of coiled DNA containing many genes, regulatory elements and other nucleotide sequences. Chromosomes also contain DNA-bound proteins, which serve to package the DNA and control its functions.

Chromosomes vary widely between different organisms. The DNA molecule may be circular or linear, and can be composed of 100,000 to 10,000,000,000 nucleotides in a long chain. Typically, eukaryotic cells (cells with nuclei) have large linear chromosomes and prokaryotic cells (cells without defined nuclei) have smaller circular chromosomes, although there are many exceptions to this rule. Also, cells may contain more than one type of chromosome; for example, mitochondria in most eukaryotes and chloroplasts in plants have their own small chromosomes. In eukaryotes, nuclear chromosomes are packaged by proteins into a condensed structure called chromatin. This allows the very long DNA molecules to fit into the cell nucleus. The structure of chromosomes and chromatin varies through the cell cycle. Chromosomes are the essential unit for cellular division and must be replicated, divided, and passed successfully to their daughter cells so as to ensure the genetic diversity and survival of their progeny. Chromosomes may exist as either duplicated or unduplicated. Unduplicated chromosomes are single linear strands, whereas duplicated chromosomes contain two identical copies (called chromatids) joined by a centromere. Compaction of the duplicated chromosomes during mitosis and meiosis results in the classic four-arm structure. Chromosomal recombination plays a vital role in genetic diversity. If these structures are manipulated incorrectly, through processes known as chromosomal instability and translocation, the cell may undergo mitotic catastrophe and die, or it may unexpectedly evade apoptosis leading to the progression of cancer. In practice "chromosome" is a rather loosely defined term.

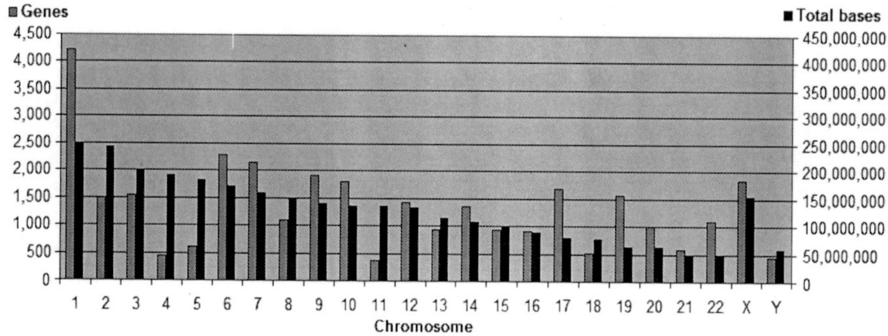

In prokaryotes and viruses, the term genophore is more appropriate when no chromatin is present. However, a large body of work uses the term chromosome regardless of chromatin content. In prokaryotes, DNA is usually arranged as a circle, which is tightly coiled in on itself, sometimes accompanied by one or more smaller, circular DNA molecules called plasmids. These small circular genomes are also found in mitochondria and chloroplasts, reflecting their bacterial origins. The simplest genophores are found in viruses: these DNA or RNA molecules are short linear or circular genophores that often lack structural proteins.

The word *chromosome* comes from the Greek *chroma*, colour and *soma*, body due to their property of being very strongly stained by particular dyes.

In a series of experiments beginning in the mid-1880s, Theodor Boveri gave the definitive demonstration that chromosomes are the vectors of heredity. His two principles were the *continuity* of chromosomes and the *individuality* of chromosomes. It is the second of these principles that was so original.

Wilhelm Roux suggested that each chromosome carries a different genetic load. Boveri was able to test and confirm this hypothesis. Aided by the rediscovery at the start of the 1900s of Gregor Mendel's earlier work, Boveri was able to point out the connection between the rules of inheritance and the behaviour of the chromosomes. Boveri influenced two generations of American cytologists: Edmund Beecher Wilson, Walter Sutton and Theophilus Painter were all influenced by Boveri (Wilson and Painter actually worked with him).

In his famous textbook *The Cell in Development and Heredity*, Wilson linked together the independent work of Boveri and Sutton (both around 1902) by naming the chromosome theory of inheritance the "Sutton-Boveri Theory" (the names are sometimes reversed).

Ernst Mayr remarks that the theory was hotly contested by some famous geneticists: William Bateson, Wilhelm Johannsen, Richard Goldschmidt and T.H. Morgan, all of a rather dogmatic turn of mind. Eventually, complete proof came from chromosome maps in Morgan's own lab.

In Eukaryotes

Eukaryotes (cells with nuclei such as those found in plants, yeast, and animals) possess multiple large linear chromosomes contained in the cell's nucleus. Each chromosome has one centromere, with one or two arms projecting from the centromere, although, under most circumstances, these arms are not visible as such. In addition, most eukaryotes have a small circular mitochondrial genome, and some eukaryotes may have additional small circular or linear cytoplasmic chromosomes. In the nuclear chromosomes of eukaryotes, the uncondensed DNA exists in a semi-ordered structure, where it is wrapped around histones (structural proteins), forming a composite material called chromatin.

Chromatin

Chromatin is the complex of DNA and protein found in the eukaryotic nucleus, which packages chromosomes. The structure of chromatin varies significantly between different stages of the cell cycle, according to the requirements of the DNA.

Interphase Chromatin

During interphase (the period of the cell cycle where the cell is not dividing), two types of chromatin can be distinguished:

- Euchromatin, which consists of DNA that is active, e.g., being expressed as protein.
- Heterochromatin, which consists of mostly inactive DNA. It seems to serve structural purposes during the chromosomal stages. Heterochromatin can be further distinguished into two types:
  - *Constitutive heterochromatin*, which is never expressed. It is located around the centromere and usually contains repetitive sequences.
  - *Facultative heterochromatin*, which is sometimes expressed.

Individual chromosomes cannot be distinguished at this stage – they appear in the nucleus as a homogeneous tangled mix of DNA and protein.

Metaphase Chromatin and Division

In the early stages of mitosis or meiosis (cell division), the chromatin strands become more and more condensed. They cease to function as accessible genetic material (transcription stops) and become a compact transportable form. This compact form makes the individual chromosomes visible, and they form the classic four arm structure, a pair of sister chromatids attached to each other at the centromere. The shorter arms are called *p arms* (from the French *petit*, small) and the longer arms are called *q arms* (*q* follows *p* in the Latin alphabet; q-g "grande"). This is the only natural context in which individual chromosomes are visible with an optical microscope.

During mitosis, microtubules grow from centrosomes located at opposite ends of the cell and also attach to the centromere at specialised structures called kinetochores, one of which is present on each sister chromatid. A special DNA base sequence in the region of the kinetochores provides, along with special proteins, longer-lasting attachment in this region. The microtubules then pull the chromatids apart towards the centrosomes, so that each daughter cell inherits one set of chromatids. Once the cells have divided, the chromatids are uncoiled and DNA can again be transcribed. In spite of their appearance, chromosomes are structurally highly condensed, which enables these giant DNA structures to be contained within a cell nucleus.

Table : *Estimated number of genes and base pairs on each human chromosome. The right-hand legend of this chart is mislabelled "Total bases". It should say "Total base* pairs".

Chromosome	*Genes*	*Total base pairs*	*Sequenced base pairs*
1	4,220	247,199,719	224,999,719
2	1,491	242,751,149	237,712,649
3	1,550	199,446,827	194,704,827
4	446	191,263,063	187,297,063
5	609	180,837,866	177,702,766
6	2,281	170,896,993	167,273,993
7	2,135	158,821,424	154,952,424
8	1,106	146,274,826	142,612,826
9	1,920	140,442,298	120,312,298
10	1,793	135,374,737	131,624,737
11	379	134,452,384	131,130,853
12	1,430	132,289,534	130,303,534
13	924	114,127,980	95,559,980
14	1,347	106,360,585	88,290,585
15	921	100,338,915	81,341,915
16	909	88,822,254	78,884,754
17	1,672	78,654,742	77,800,220
18	519	76,117,153	74,656,155
19	1,555	63,806,651	55,785,651
20	1,008	62,435,965	59,505,254
21	578	46,944,323	34,171,998
22	1,092	49,528,953	34,893,953
X (sex chromosome)	1,846	154,913,754	151,058,754
Y (sex chromosome)	454	57,741,652	25,121,652
Total	32,185	3,079,843,747	2,857,698,560

In Prokaryotes

The prokaryotes – bacteria and archaea – typically have a single circular chromosome, but many variations exist. Most bacteria have a single circular chromosome that can range in size from only 160,000 base pairs in the endosymbiotic bacterium *Candidatus Carsonella ruddii*, to 12,200,000 base pairs in the soil-dwelling bacterium *Sorangium cellulosum*. Spirochaetes of the genus *Borrelia* are a notable exception to this arrangement, with bacteria such as *Borrelia burgdorferi*, the cause of Lyme disease, containing a single linear chromosome.

Structure in Sequences

Prokaryotic chromosomes have less sequence-based structure than eukaryotes. Bacteria typically have a single point (the origin of replication) from which replication starts, whereas some archaea contain multiple replication origins. The genes in prokaryotes are often organised in operons, and do not usually contain introns, unlike eukaryotes.

DNA Packaging

Prokaryotes do not possess nuclei. Instead, their DNA is organised into a structure called the nucleoid. The nucleoid is a distinct structure and occupies a defined region of the bacterial cell. This structure is, however, dynamic and is maintained and remodelled by the actions of a range of histone-like proteins, which associate with the bacterial chromosome. In archaea, the DNA in chromosomes is even more organised, with the DNA packaged within structures similar to eukaryotic nucleosomes.

Bacterial chromosomes tend to be tethered to the plasma membrane of the bacteria. In molecular biology application, this allows for its isolation from plasmid DNA by centrifugation of lysed bacteria and pelleting of the membranes (and the attached DNA).

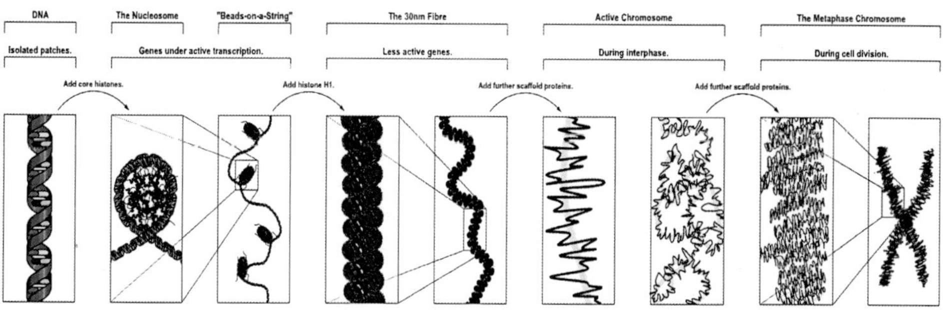

Prokaryotic chromosomes and plasmids are, like eukaryotic DNA, generally supercoiled. The DNA must first be released into its relaxed state for access for transcription, regulation, and replication.

Number of Chromosomes in Various Organisms

Eukaryotes

For example, human cells are diploid and have 22 different types of autosome, each present as two copies, and two sex chromosomes. This gives 46 chromosomes in total. Other organisms have more than two copies of their chromosomes, such as bread wheat, which is *hexaploid* and has six copies of seven different chromosomes – 42 chromosomes in total. Normal members of a particular eukaryotic species all have the same number of nuclear chromosomes. Other eukaryotic chromosomes, i.e., mitochondrial and plasmid-like small chromosomes, are much more variable in number, and there may be thousands of copies per cell. Asexually reproducing species have one set of chromosomes, which are the same in all body cells. However, asexual species can be either haploid or diploid.

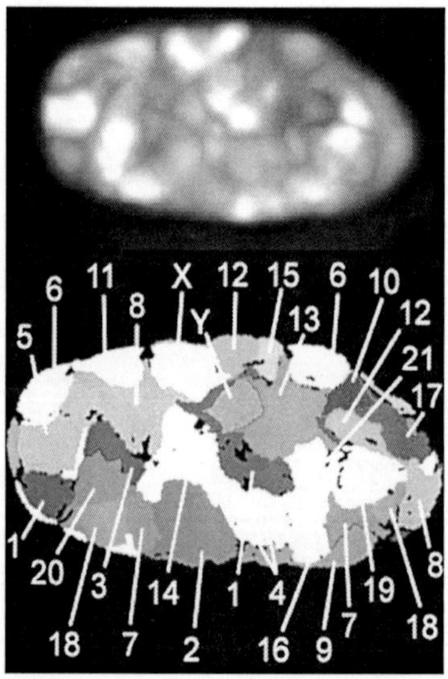

Figure: *The 23 human chromosome territories during prometaphase in fibroblast cells.*

Sexually reproducing species have somatic cells (body cells), which are diploid [2n] having two sets of chromosomes, one from the mother and one from the father. Gametes, reproductive cells, are haploid [n]: They have one set of chromosomes. Gametes are produced by meiosis of a diploid germ line cell. During meiosis, the matching chromosomes of father and mother can exchange small parts of themselves (crossover), and thus create new chromosomes that are not inherited solely from either parent. When a male and a female gamete merge (fertilization), a new diploid organism is formed.

Some animal and plant species are polyploid [Xn]: They have more than two sets of homologous chromosomes. Plants important in agriculture such as tobacco or wheat are often polyploid, compared to their ancestral species.

Wheat has a haploid number of seven chromosomes, still seen in some cultivars as well as the wild progenitors. The more-common pasta and bread wheats are polyploid, having 28 (tetraploid) and 42 (hexaploid) chromosomes, compared to the 14 (diploid) chromosomes in the wild wheat.

Prokaryotes

Prokaryote species generally have one copy of each major chromosome, but most cells can easily survive with multiple copies. For example, *Buchnera*, a symbiont of aphids has multiple copies of its chromosome, ranging from 10–400 copies per cell. However, in some large bacteria, such as *Epulopiscium fishelsoni* up to 100,000 copies of the chromosome can be present. Plasmids and plasmid-like small chromosomes are, as in eukaryotes, very variable in copy number. The number of plasmids in the cell is almost entirely determined by the rate of division of the plasmid – fast division causes high copy number.

Karyotype

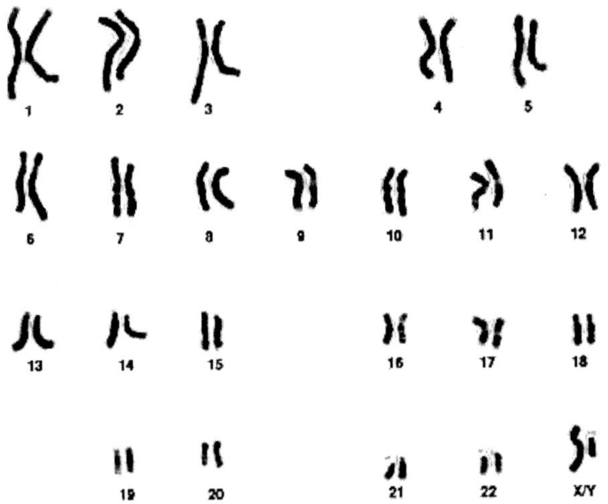

Figure: *Karyogram of a human male*

In general, the karyotype is the characteristic chromosome complement of a eukaryote species. The preparation and study of karyotypes is part of cytogenetics. Although the replication and transcription of DNA is highly standardised in eukaryotes, *the same cannot be said for their karyotypes*, which are often highly variable. There may be variation between species in chromosome number and in detailed organisation. In some cases, there is significant variation within species. Often there is:

1. Variation between the two sexes
2. Variation between the germ-line and soma (between gametes and the rest of the body)

3. Variation between members of a population, due to balanced genetic polymorphism
4. Geographical variation between races
5. Mosaics or otherwise abnormal individuals.

Also, variation in karyotype may occur during development from the fertilized egg. The technique of determining the karyotype is usually called *karyotyping*. Cells can be locked part-way through division (in metaphase) in vitro (in a reaction vial) with colchicine. These cells are then stained, photographed, and arranged into a *karyogram*, with the set of chromosomes arranged, autosomes in order of length, and sex chromosomes (here X/Y) at the end. Like many sexually reproducing species, humans have special gonosomes (sex chromosomes, in contrast to autosomes). These are XX in females and XY in males.

Historical Note

Investigation into the human karyotype took many years to settle the most basic question: *How many chromosomes does a normal diploid human cell contain?* In 1912, Hans von Winiwarter reported 47 chromosomes in spermatogonia and 48 in oogonia, concluding an XX/XO sex determination mechanism.

Painter in 1922 was not certain whether the diploid number of man is 46 or 48, at first favouring 46. He revised his opinion later from 46 to 48, and he correctly insisted on humans having an XX/XY system.

New techniques were needed to definitively solve the problem:

1. Using cells in culture
2. Pretreating cells in a hypotonic solution, which swells them and spreads the chromosomes
3. Arresting mitosis in metaphase by a solution of colchicine
4. Squashing the preparation on the slide forcing the chromosomes into a single plane
5. Cutting up a photomicrograph and arranging the result into an indisputable karyogram.

It took until 1954 before the human diploid number was confirmed as 46. Considering the techniques of Winiwarter and Painter, their results were quite remarkable. Chimpanzees (the closest living relatives to modern humans) have 48 chromosomes.

Aberrations

Chromosomal aberrations are disruptions in the normal chromosomal content of a cell and are a major cause of genetic conditions in humans, such as Down syndrome, although most aberrations have little to no effect. Some chromosome abnormalities do not cause disease in carriers, such as translocations, or chromosomal inversions, although they may lead to a higher chance of birthing a child with a chromosome disorder.

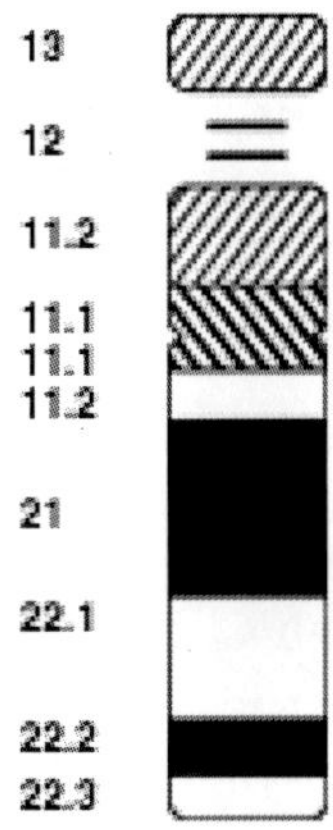

Figure: *In Down syndrome, there are three copies of chromosome 21*

Abnormal numbers of chromosomes or chromosome sets, called aneuploidy, may be lethal or may give rise to genetic disorders. Genetic counselling is offered for families that may carry a chromosome rearrangement. The gain or loss of DNA from chromosomes can lead to a variety of genetic disorders. Human examples include:

- Cri du chat, which is caused by the deletion of part of the short arm of chromosome 5. "Cri du chat" means "cry of the cat" in French; the condition was so-named because affected babies make high-pitched cries that sound like those of a cat. Affected individuals have wide-set eyes, a small head and jaw, moderate to severe mental health issues, and are very short.
- Down syndrome, the most common trisomy, usually caused by an extra copy of chromosome 21 (trisomy 21). Characteristics include decreased muscle tone, stockier build, asymmetrical skull, slanting eyes and mild to moderate developmental disability.
- Edwards syndrome, or trisomy-18, the second-most-common trisomy. Symptoms include motor retardation, developmental disability and numerous congenital anomalies causing serious health problems. Ninety percent of those affected die in infancy; however, those that live past their first birthday are usually quite healthy thereafter. They have characteristic clenched hands and overlapping fingers.
- Isodicentric 15, also called idic(15), partial tetrasomy 15q, or inverted duplication 15 (inv dup 15).

- Jacobsen syndrome, which is very rare. It is also called the terminal 11q deletion disorder. Those affected have normal intelligence or mild developmental disability, with poor expressive language skills. Most have a bleeding disorder called Paris-Trousseau syndrome.
- Klinefelter's syndrome (XXY): Men with Klinefelter syndrome are usually sterile, and tend to be taller and have longer arms and legs than their peers. Boys with the syndrome are often shy and quiet, and have a higher incidence of speech delay and dyslexia. Without testosterone treatment, some may develop gynecomastia during puberty.
- Patau Syndrome, also called D-Syndrome or trisomy-13. Symptoms are somewhat similar to those of trisomy-18, without the characteristic folded hand.
- Small supernumerary marker chromosome: This means there is an extra, abnormal chromosome. Features depend on the origin of the extra genetic material. Cat-eye syndrome and isodicentric chromosome 15 syndrome (or Idic15) are both caused by a supernumerary marker chromosome, as is Pallister-Killian syndrome.
- Triple-X syndrome (XXX): XXX girls tend to be tall and thin and have a higher incidence of dyslexia.
- Turner syndrome (X instead of XX or XY): In Turner syndrome, female sexual characteristics are present but underdeveloped. Females with Turner syndrome often have a short stature, low hairline, abnormal eye features and bone development and a "caved-in" appearance to the chest.
- XYY syndrome: XYY boys are usually taller than their siblings. Like XXY boys and XXX girls, they are more likely to have learning difficulties.
- Wolf-Hirschhorn syndrome, which is caused by partial deletion of the short arm of chromosome 4. It is characterised by severe growth retardation and severe to profound mental health issues.

Genetic Deletion

In genetics, a deletion also called gene deletion, deficiency, or deletion mutation) is a mutation (a genetic aberration) in which a part of a chromosome or a sequence of DNA is missing. Deletion is the loss of genetic material. Any number of nucleotides can be deleted, from a single base to an entire piece of chromosome. Deletions can be caused by errors in chromosomal crossover during meiosis. This causes several serious genetic diseases. Deletion is also causing frameshift.

Causes

Causes include the following:

- Losses from translocation

- Chromosomal crossovers within a chromosomal inversion
- Unequal crossing over
- Breaking without rejoining.

For synapsis to occur between a chromosome with a large intercalary deficiency and a normal complete homolog, the unpaired region of the normal homolog must loop out of the linear structure into a deletion or compensation loop.

Types

Types of deletion include the following:

- Terminal Deletion - a deletion that occurs towards the end of a chromosome.
- Intercalary Deletion / Interstitial Deletion - a deletion that occurs from the interior of a chromosome.

Effects

Small deletions are less likely to be fatal; large deletions are usually fatal – there are always variations based on which genes are lost. Some medium-sized deletions lead to recognizable human disorders. Deletion of a number of base pairs that is not evenly divisible by three will lead to a frameshift mutation, causing all of the codons occurring after the deletion to be read incorrectly during translation, producing a severely altered and potentially nonfunctional protein. In contrast, a deletion that is evenly divisible by three is called an *in-frame* deletion. Deletions are responsible for an array of genetic disorders, including some cases of male infertility and two thirds of cases of Duchenne muscular dystrophy. Deletion of part of the short arm of chromosome 5 results in a syndrome called Cri du chat, French for “cry of the cat” syndrome. It is found in approximately 1 in 50,000 live births. The surviving infants have a distinctive cry, severe mental retardation, and shortened life span.

Detection

The introduction of molecular techniques in conjunction with classical cytogenetic methods has in recent years greatly improved the diagnostic potential for chromosomal abnormalities. In particular, microarray-comparative genomic hybridization (CGH) based on the use of BAC clones promises a sensitive strategy for the detection of DNA copy-number changes on a genome-wide scale. The resolution of detection could as high as >30,000 “bands” and the size of chromosomal deletion detected could as small as 5–20 kb in length.

Chromosome (Genetic Algorithm)

In genetic algorithms, a chromosome (also sometimes called a genome) is a set of parameters which define a proposed solution to the problem that the genetic algorithm is trying to solve. The chromosome is often represented as a simple string, although a wide variety of other data structures are also used.

Chromosome Design

The design of the chromosome and its parameters is by necessity specific to the problem to be solved. To give a trivial example, suppose the problem is to find the integer value of x between 0 and 255 that provides the maximal result for $f(x) = x^2$. (This is not the type of problem that is normally solved by a genetic algorithm, since it can be trivially solved using numeric methods. It is only used to serve as a simple example.) Our possible solutions are the integers from 0 to 255, which can all be represented as 8 digit binary strings. Thus, we might use an 8 digit binary string as our chromosome. If a given chromosome in the population represents the value 155, its chromosome would be 10011011. A more realistic problem we might wish to solve is the travelling salesman problem. In this problem, we seek an ordered list of cities that results in the shortest trip for the salesman to travel. Suppose there are six cities, which we'll call A, B, C, D, E, and F. A good design for our chromosome might be the ordered list we want to try. An example chromosome we might encounter in the population might be DFABEC.

The mutation operator and crossover operator employed by the genetic algorithm must take into account the chromosome's design.

- Genetic genealogy
  - o Genealogical DNA test
- List of number of chromosomes of various organisms
- Locus (explains gene location nomenclature)
- Maternal influence on sex determination
- Sex-determination system
- o XY sex-determination system
  - * X-chromosome
  - * X-inactivation
  - * Y-chromosome
  - * Y-chromosomal Aaron
  - * Y-chromosomal Adam
- Lampbrush chromosome
- Polytene chromosome.

Genes in Pedigrees

The simplest genetic characters are those whose presence or absence depends on the genotype at a single locus. That is not to say that the character itself is programmed by only one pair of genes - expression of any human character is likely to require a large number of genes and environmental factors.

However, sometimes a particular genotype at one locus is both necessary and sufficient for the character to be expressed, given the normal genetic and environmental background of the organism. Such characters are called mendelian. In humans over 10,000 mendelian characters are known.

Dominance and Recessiveness are Properties of Characters, not Genes

A character is dominant if it is manifest in the heterozygote and recessive if not. Note that dominance and recessiveness are properties of characters, not genes. Thus sickle cell anemia is recessive because only HbS homozygotes manifest it. Heterozygotes for the same gene show sickling trait, which is therefore a dominant character. Most human dominant syndromes are known only in heterozygotes.

Sometimes homozygotes have been described, born from matings of two heterozygous affected people, and often the homozygotes are much more severely affected. Examples are achondroplasia (short-limbed dwarfism) and Type 1 Waardenburg syndrome (deafness with pigmentary abnormalities).

Nevertheless we describe achondroplasia and Waardenburg syndrome as dominant because these terms describe phenotypes seen in heterozygotes. In experimental organisms, where this uncertainty does not exist, geneticists tend to use the term semidominant when the heterozygote has an intermediate phenotype, reserving 'dominant' for conditions where the homozygote is indistinguishable from the heterozygote - Huntington disease (adult-onset progressive neurological deterioration) for example.

The question of dominance has been well reviewed by Wilkie (1994). Males are hemizygous for loci on the X and Y chromosomes, where they have only a single copy of each gene, so the question of dominance or recessiveness does not arise in males for X- or Y-linked characters.

X-inactivation (Lyonisation) Blurs the Distinction between Dominant and Recessive X-linked Conditions

Carriers of 'recessive' X-linked conditions often manifest some signs, while compared to affected males, heterozygotes for 'dominant' conditions are usually more mildly and variably affected. This is a consequence of X-inactivation. Mammals compensate for the unequal numbers of X chromosomes in male and female cells by permanently inactivating all but one X chromosome in each cell. XY males keep their single X active, whilst XX females inactivate one X (chosen at random) in each cell. Inactivation takes place early in embryonic life, and once a cell has chosen which X to inactivate, that choice is transmitted clonally to all its daughter cells. A female heterozygous for an X-linked condition (dominant or recessive), is a mosaic. Each cell expresses either the normal or the abnormal allele, but not both. Where the phenotype depends on a circulating product, as in hemophilia (failure of blood to clot), there is an averaging effect between the normal and abnormal cells. Female carriers have an intermediate phenotype, and are usually clinically unaffected but biochemically abnormal. Where the phenotype is a localized property of individual cells, as in hypohidrotic ectodermal dysplasia (MIM 305100: missing sweat glands, abnormal teeth and hair) female carriers show patches of normal and abnormal tissue. Occasional manifesting heterozygotes are seen for X-linked recessive conditions: these women may be quite severely affected because by bad luck most cells in some critical tissue have inactivated the normal X.

There are Probably no Y-linked Diseases

Although a few Y-linked characters have been described, no Y-linked diseases are known, apart from disorders of male sexual function. Conceivably such a disease may exist undiscovered, but this is unlikely for two reasons. First, the pedigree pattern would be strikingly noticeable, especially in societies that trace family through the male line, yet they have not been noted. Second, the Y-chromosome cannot carry any genes whose function is important for health, because females are perfectly normal without any Y-linked genes. Thus any Y-linked genes must code either for non-essential characters or for male-specific functions, and defects are unlikely to cause diseases apart from defects of male sexual function. Genes present as functional copies on both the Y and the X might prove an exception to this argument.

The Mode of Inheritance can Rarely be Defined Unambiguously in a Single Pedigree

Given the limited size of human families, it is rarely possible to be completely certain of the mode of inheritance of a character simply by inspecting a single pedigree. In experimental animals one would set up a test cross and check for a 1 in 2 or 1 in 4 ratio. In human pedigrees the proportion of affected children is not a very reliable indicator. Mostly this is because the numbers are too small, but in addition, the way in which the family was ascertained can bias the ratio of affected to unaffected children observed. For recessive conditions, the proportion of affected children often seems to be greater than 1 in 4. This is because families are normally ascertained when they have an affected child; families where both parents are carriers but, by good fortune, nobody is affected are systematically missed.

For many of the rarer conditions, the stated mode of inheritance is no more than an informed guess. Assigning modes of inheritance is important, because that is the basis of the risk estimates used in genetic counselling. However, it is important to recognize that the modes of inheritance are often working hypotheses rather than established fact. OMIM uses an asterisk to denote entries with relatively well-established modes of inheritance. Only when a cloned copy of the gene is available can the inheritance be determined with certainty.

One Gene - one Enzyme Does not Imply one gene - one Syndrome

Pedigree patterns provide the essential entry point into human genetics, but they are only a starting point for defining genes. It would be a serious error to imagine that the 10,000 or so known mendelian characters define 10,000 DNA coding sequences. This would be an unjustified extension of the one gene - one enzyme hypothesis of Beadle and Tatum.

Back in the 1940s this hypothesis allowed a major leap forward in understanding how genes determine phenotypes. Since then it has been extended - some genes encode nontranslated RNAs, some proteins are not enzymes, and many proteins contain several separately encoded polypeptide chains. But even with these extensions, Beadle and Tatum's hypothesis cannot be used to imply a one-to-one correspondence between entries in the OMIM catalogue and entries in Genbank, the DNA sequence database.

The genes of classical genetics are abstract entities. Any character that is determined at a single chromosomal location will segregate in a mendelian pattern - but the determinant may not be a gene in the molecular geneticist's sense of the word. Fascio-scapulo-humeral muscular dystrophy (MIM 158900: severe but nonlethal weak ness of certain muscle groups) is associated with small deletions of sequences at 4q35, but nobody (at the time of writing) has managed to find a protein-coding sequence at that location, despite intensive searching and sequencing. The 'gene' for Charcot-Marie-Tooth disease type 1A (MIM 118220: motor and sensory neuropathy) turned out to be a 1.5 Mb tandem duplication on chromosome 17. These examples are unusual; most OMIM entries probably do describe the consequences of mutations affecting a single transcription unit, but because of locus and allelic heterogeneity, there is still no one-to-one correspondence with Genbank entries.

Locus Heterogeneity is Common in Syndromes That Result from Failure of a Complex Pathway

Profound congenital hearing loss is often genetic, and when genetic it is usually autosomal recessive. However, when two people with autosomal recessive profound hearing loss marry, as they often do, the children usually have normal hearing. This is an example of complementation. The children will have normal hearing whenever the parents carry mutations in different genes. Diseases and developmental defects represent the failure of a pathway.

It is easy to see that many different genes would be needed to construct so exquisite a machine as the cochlear hair cell, and a defect in any of those genes could lead to deafness. Such locus heterogeneity is only to be expected in conditions like deafness, blindness or mental retardation, where a rather general pathway has failed; but even with more specific pathologies, multiple loci are very frequent. A striking example is Usher syndrome, an autosomal recessive combination of hearing loss and retinitis pigmentosa, which can be caused by mutations at eight or more unlinked loci (Smith et al., 1994). OMIM has separate entries for known examples of locus heterogeneity (defined by linkage or mutation analysis), but there must be many undetected examples still contained within single entries.

Allelic Series are a Cause of Clinical Heterogeneity

Sometimes several apparently distinct human phenotypes turn out to be all caused by different allelic mutations at the same locus. The difference may be one of degree - mutations that partially inactivate the dystrophin gene produce Becker muscular dystrophy, while mutations that completely inactivate the same gene produce the similar but more severe Duchenne muscular dystrophy (lethal muscle wasting). Sometimes the difference is qualitative - inactivation of the androgen receptor gene causes androgen insensitivity (MIM 313700; 46,XY embryos develop as females), but expansion of a run of glutamine codons within the same gene causes a very different disease, spinobulbar muscular atrophy or Kennedy disease (MIM 313200).

Mitochondrial Inheritance Gives a Recognizable Matrilinear Pedigree Pattern

In addition to the mutations in genes carried on the nuclear chromosomes, mitochondrial mutations are a significant cause of human genetic disease. The mitochondrial genome is small but highly mutable compared to nuclear DNA, probably because mitochondrial DNA replication is more error-prone and the number of replications is much higher. Mitochondrially-encoded diseases have two unusual features, matrilineal inheritance and frequent heteroplasmy (Wallace, 1994). Inheritance is matrilineal, because sperm do not contribute mitochondria to the zygote (this assertion rests on limited evidence; however, paternally-derived mitochondrial variants are not detected in children). Thus a mitochondrially inherited condition can affect both sexes, but is passed on only by affected mothers, giving a recognizable pedigree pattern.

Complications to the Basic Pedigree Patterns

Common Recessive Conditions can Give a Pseudo-dominant Pedigree Pattern

If a character is common in the population, there is a high chance that it may be brought into the pedigree independently by two or more people. A common recessive character like blood group O may be seen in successive generations because of repeated marriages of group O people with heterozygotes. This produces a pattern resembling dominant inheritance. Thus the classic pedigree patterns are best seen with rare characters.

Failure of a Dominant Condition to Manifest is Called Nonpenetrance

With dominant conditions, nonpenetrance is a frequent complication. The penetrance of a character, for a given genotype, is defined as the probability that a person who has the genotype will manifest the character. By definition, a dominant character is manifest in a heterozygous person, and so should show 100% penetrance. Nevertheless, many human characters, while generally showing dominant inheritance, occasionally skip a generation.

There is no mystery about nonpenetrance; indeed, 100% penetrance is the more surprising phenomenon. Very often the presence or absence of a character depends, in the main and in normal circumstances, on the genotype at one locus, but an unusual genetic background, a particular lifestyle or maybe just chance means that the occasional person may fail to manifest the character. Nonpenetrance is a major pitfall in genetic counselling. One of the jobs of genetic counselors is to know the usual degree of penetrance of each dominant syndrome. Frequently, of course, a character depends on many factors and does not show a mendelian pedigree pattern even if entirely genetic. There is a continuum of characters from fully penetrant mendelian to multifactorial, with increasing influence of other genetic loci and/or the environment. No logical break separates imperfectly penetrant mendelian from multifactorial characters; it is a question of which is the most useful description to apply.

Late-onset Diseases Show Age-related Penetrance

A particularly important case of reduced penetrance is seen with late-onset diseases. Genetic conditions are, of course, not necessarily congenital (present at birth). The genotype is fixed at conception, but the phenotype may not manifest until adult life. In such cases the penetrance is age-related. Huntington disease is a well-known example. Delayed onset might be caused by slow accumulation of a noxious substance, by slow tissue death or by inability to repair some form of environmental damage. Hereditary cancers are caused by a second mutation affecting a cell of a person who already carries one mutation in a tumour suppressor gene. Depending on the disease, the penetrance may become 100% if the person lives long enough, or there may be people who carry the gene but who will never develop symptoms no matter how long they live.

Many Conditions Show Variable Expression

Different family members show different features of the syndrome. The cause is the same as with nonpenetrance: other genes, environmental factors or pure chance have some influence on development of the symptoms. Nonpenetrance and variable expression are typically problems with dominant, rather than recessive, characters. Partly this reflects the difficulty of spotting nonpenetrant cases in a typical recessive pedigree. However, as a general rule, recessive conditions are less variable than dominant ones, probably because the phenotype of a heterozygote involves a balance between the effects of the two alleles, so that the outcome is likely to be more sensitive to outside influence than the phenotype of a homozygote. However, both nonpenetrance and variable expression are occasionally seen in recessive conditions. These complications are much more conspicuous in humans than in plants or other animals, because laboratory animals and crop plants are far more genetically uniform than humans. What we see in human genetics is typical of a wild population. Nevertheless, mouse geneticists are familiar with the way expression of a mutant can change when it is bred onto a different genetic background, and understand its importance when studying mouse models of human diseases.

Anticipation is a Special Type of Variable Expression

Anticipation describes the tendency of some variable dominant conditions to become more severe in successive generations. Until recently, most geneticists were skeptical that this ever really happened. The problem is that true anticipation is very easily mimicked by random variations in severity. A family comes to clinical attention when a severely affected child is born. Investigating the history, the geneticist notes that one of the parents is affected, but only mildly. This looks like anticipation, but may actually be just a bias of ascertainment. Had the parent been severely affected, he or she would most likely never have become a parent, and had the child been mildly affected, the family would not have come to notice. Given the lack of any plausible mechanism for anticipation, and the statistical problems of demonstrating it in the face of these biases, most geneticists were unwilling to consider anticipation seriously until molecular developments obliged them to do so.

Anticipation suddenly became respectable, even fashionable, with the discovery of unstable expanding trinucleotide repeats in Fragile-X syndrome (MIM 309550: mental retardation with various physical signs), and later in myotonic dystrophy (MIM 160900: a very variable multisystem disease with characteristic muscular dysfunction) and Huntington disease. Severity or age of onset of these diseases correlates with the repeat length, and the repeat length tends to grow as the gene is transmitted down the generations. Thus these conditions show true anticipation. Now once again we see claims for anticipation being made for many diseases, and it is important to bear in mind that the old objection about bias of ascertainment remains valid. To be credible, a claim of anticipation requires careful statistical backing, and not just anecdotal evidence.

For Imprinted Genes, Expression Depends on Parental Origin

Certain human characters are autosomal dominant and transmitted by parents of either sex, but they manifest only when inherited from a parent of one particular sex. For example there are families with autosomal dominant glomus tumours that are expressed only in people who inherit the gene from their father, while Beckwith-Wiedemann syndrome (MIM 130650: exomphalos, macroglossia, overgrowth) is sometimes dominant but expressed only by people who inherit it from their mother. These parental sex effects are evidence of imprinting, a poorly understood phenomenon whereby certain genes are somehow marked (imprinted) with their parental origin.

Male Lethality may Complicate X-linked Pedigrees

For some X-linked dominant conditions, absence of the normal allele is lethal before birth. Thus affected males are not born, and we see a condition that affects only females, who pass it on to half their daughters but none of their sons. There may be a history of miscarriages, but families are rarely big enough to prove that the number of sons is only half the number of daughters. An example is incontinentia pigmenti (MIM 308310: linear skin defects following defined patterns known as Blaschko's lines, often accompanied by neurological or skeletal problems).

New Mutations Often Complicate Pedigree Interpretation, and can Lead to Mosaicism

Many cases of severe genetic disease are the result of fresh mutations, striking without warning in a family with no previous history of the disease. People with severe genetic diseases seldom reproduce, so they do not pass on their mutant genes. On the assumption that, averaged over time, new mutations exactly replace the disease genes lost through natural selection, there is a simple relationship between the rate at which natural selection is removing disadvantageous genes, the rate at which new mutation is creating them, and their frequency in the population.

This mutation-selection dynamic has different effects on pedigrees, depending on the mode of inheritance. Autosomal recessive pedigrees are not significantly affected - any new mutations probably happened many generations ago, and we can safely assume that the parents of an affected child are both carriers. For dominant conditions however the turnover of disease genes is much faster, because they are constantly exposed to selection. A fully penetrant lethal dominant would necessarily always occur by fresh mutation, and the parents would never be affected (an example is thanatophoric dysplasia, MIM 187600: severe shortening of long bones and abnormal fusion of cranial sutures). People with nonlethal but severe dominant conditions often have unaffected parents and no previous family history of the condition. Serious X-linked recessives also show a significant proportion of fresh mutations, because the gene is exposed to natural selection whenever it is in a male. When a normal couple with no relevant family history have a child with severe abnormalities, deciding the mode of inheritance and recurrence risk can be very difficult: the problem might be autosomal recessive, autosomal dominant with a new mutation, X-linked recessive (if the child is male) or nongenetic. A further complication is introduced by germinal mosaicism.

Mosaics Have Two (Or more) Genetically Different Cell Lines

We have seen that in serious autosomal dominant and X-linked diseases, where affected people have few or no children, the disease genes are maintained in the population by recurrent mutation. A common assumption is that an entirely normal person produces a single mutant gamete. However, this is not necessarily what happens. Unless there is something special about the mutational process, such that it can happen only during gametogenesis, mutations may arise at any time during post-zygotic life. Post-zygotic mutations produce mosaics with two (or more) genetically distinct cell lines. The older literature on human mosaicism refers only to chromosomal mosaicism, because that was the only type of mosaicism that could be detected before DNA analysis was developed, but mosaicism for single gene mutations is at least as frequent and important.

Mosaicism can affect somatic and/or germ line tissues. Post-zygotic mutations are not merely frequent, they are inevitable. Human mutation rates are typically 10^{-7} per gene per cell generation, and our bodies contain perhaps 10^{13} cells. It follows that every one of us must be a mosaic for innumerable genetic diseases. Indeed, as Professor John Edwards memorably remarked, a normal man may well produce the whole of the OMIM catalogue in every ejaculate. This should cause no anxiety. If a cell in your finger mutates to the Huntington disease genotype, or a cell in your ear picks up a cystic fibrosis mutation, there are absolutely no consequences for you or your family. Only if a somatic mutation results in the emergence of a substantial clone of mutant cells is there a risk to the whole organism. This can happen in two ways:

1. The mutation causes abnormal proliferation of a cell that would normally replicate little or not at all, thus generating a clone of mutant cells.
2. the mutation occurs in an early embryo, affecting a cell which is the progenitor of a significant fraction of the whole organism. In that case the mosaic individual may show clinical signs of disease.

Mutations occurring in a parent's germ line can cause *de novo* inherited disease in a child. When an early germ-line mutation has produced a person who harbours a large clone of mutant germ-line cells (germinal, or gonadal, mosaicism), a normal couple with no previous family history may produce more than one child with the same serious dominant disease. The pedigree mimicks recessive inheritance. Even if the correct mode of inheritance is realised, it is very difficult to calculate a recurrence risk to use in counselling the parents. Usually an empiric risk is quoted.

A new mutation in X-linked recessive Duchenne muscular dystrophy. The three grandparental X chromosomes were distinguished using genetic markers, and are shown in blue, gray and white.

Molecular studies can be a great help in these cases. Sometimes it is possible to demonstrate directly that a normal father is producing a proportion of mutant sperm. Direct testing of the germ line is not possible in women, but other accessible tissues such as fibroblasts or hair roots can be examined for evidence of mosaicism. A negative result on somatic tissues does not rule out germ line mosaicism, but a positive result, in conjunction with an affected child, proves it.

Chimeras Contain Cells From two Separate Zygotes in a Single Organism

Mosaics are presumed (though rarely proved) to derive from a single fertilized egg. Chimeras on the other hand are the result of fusion of two zygotes into a single embryo (the reverse of twinning), or alternatively of limited colonisation of one twin by cells from a nonidentical co-twin. Chimerism is proved by the presence in pooled tissue samples of too many parental alleles at several loci (if just one locus were involved, one would suspect mosaicism for a single mutation). Blood-grouping centres occasionally discover chimeras among normal donors, and some intersex patients turn out to be XX/XY chimeras. Strain et al. (1998) describe a remarkable case of a 46,XY/46,XX boy whose 46,XX cell line is parthenogenetic, derived by diploidisation of a haploid maternal cell with no paternal contribution.

Factors Affecting Gene Frequencies

There can be a Simple Relation between Gene Frequencies and Genotype Frequencies

Over a whole population there may be many different alleles at a particular locus, although each individual person has just two alleles, which may be identical or different. We can imagine a gene pool, consisting of all alleles at the *A* locus in the population. The gene frequency of allele A_1 is the proportion of all *A* alleles in the gene pool which are A_1. Consider two alleles, A_1 and A_2 at the *A* locus. Let their gene frequencies be p and q respectively (p and q are each between 0 and 1). Let us perform a thought experiment:

- Pick an allele at random from the gene pool. There is a chance p that it is A_1 and a chance q that it is A_2.
- Pick a second allele at random. Again the chance of picking A_1 is p and the chance of picking A_2 is q (we assume the gene pool is sufficiently

large that removing the first allele has not significantly changed the gene frequencies of the remaining alleles).

- The chance that both alleles were A_1 is p^2.
- The chance that both alleles were A_2 is q^2.
- The chance that the first allele was A_1 and the second A_2 is pq. The chance that the first was A_2 and the second A_1 is qp. Overall, the chance of picking one A_1 and one A_2 allele is 2pq.

The Hardy-Weinberg Distribution

If we pick a person at random from the population, this is equivalent to picking two genes at random from the gene pool. The chance the person is A_1A_1 is p^2, the chance they are A_1A_2 is 2pq, and the chance they are A_2A_2 is q^2. This simple relationship between gene frequencies and genotype frequencies holds whenever a person's two genes are drawn independently and at random from the gene pool. A_1 and A_2 may be the only alleles at the locus (in which case $p + q = 1$) or there may be other alleles and other genotypes ($p + q < 1$). For X-linked loci males, being hemizygous (only one allele) are A_1 or A_2 with frequencies p and q respectively, while females can be A_1A_1, A_1A_2 or A_2A_2.

Limitations of the Hardy-Weinberg Distribution

These simple calculations break down if the underlying assumption, that a person's two genes are picked independently from the gene pool, is violated. In particular, there is a problem if there has not been random mating. Assortative mating can take several forms, but the most generally important is inbreeding. If you marry a relative you are marrying somebody whose genes resemble your own. This increases the likelihood of your children being homozygous and decreases the likelihood that they will be heterozygous. Rare recessive conditions are strongly associated with parental consanguinity, and Hardy-Weinberg calculations that ignore this will overestimate the carrier frequency in the population at large.

Use of the Hardy-Weinberg Distribution in Genetic Counselling

Gene frequencies or genotype frequencies are essential inputs into many forms of genetic analysis, such as linkage analysis and segregation analysis, and they have a particular importance in calculating genetic risks.

Genotype Frequencies can be Used (With Caution) to Calculate Mutation Rates

Mutant genes are being created by fresh mutation and being removed by natural selection. For a given level of selection we can calculate the mutation rate that would be required to replace the genes lost by selection. If we assume that there is an equilibrium in the population between the rates of loss and of replacement, the calculation tells us the present mutation rate. We can define the coefficient of selection (s) as the relative chance of reproductive failure of a genotype due to selection (the fittest type in the population has s = 0, a genetic lethal has s = 1).

- For an autosomal recessive condition, a proportion q^2 of the population are affected. The loss of disease genes each generation is sq^2. This is balanced by mutation at the rate of $\mu(1 - q^2)$ where μ is the mutation rate per gene per generation. At equilibrium $sq^2 = \mu(1 - q^2)$, or approximately (if q is small) $\mu = sq^2$.
- For a rare autosomal dominant condition homozygotes are excessively rare. Heterozygotes occur with frequency 2pq (frequency of disease gene = p). Only half the genes lost through their reproductive failure are the disease allele, so the rate of gene loss is very nearly sp. Again this is balanced by a rate of new mutation of μq^2, which is approximately μ if q is almost 1. Thus $\mu = sp$.
- For an X-linked recessive disease the rate of gene loss through affected males is sq. This is balanced by a mutation rate 3μ, since all X chromosomes in the population are available for mutation, but only the one third of X chromosomes which are in males are exposed to selection. Thus $\mu = sq/3$.

Estimates derived using them can be compared with the general expectation, from studies in many organisms, that mutation rates are typically 10^{-5}–10^{-7} per gene per generation. Mutation-selection equilibrium. μ = mutation rate per gene per generation. Heterozygote advantage can be much more important than recurrent mutation for determining the frequency of a recessive disease

The formula $\mu = sq^2$ gives an unexpectedly high mutation rate for some autosomal recessive conditions. Consider cystic fibrosis (CF), for example. Until very recently, virtually nobody with CF lived long enough to reproduce, therefore s = 1. CF affects about one birth in 2000 in the UK. Thus $q^2 = 1/2000$, and the formula gives $\mu = 5 \times 10^{-4}$. This would be a strikingly high mutation rate for any gene, but there is evidence that new CF mutations are in fact very rare. This follows from the uneven ethnic distribution of CF and the existence of strong linkage disequilibrium. The missing factor is heterozygote advantage. CF carriers have, or had in the past, some reproductive advantage over normal homozygotes. There has been much debate over what this advantage might be. The CF gene encodes a membrane chloride channel, which is required by *Salmonella typhi* for it to enter epithelial cells, so maybe heterozygotes are relatively resistant to typhoid fever (Pier et al., 1998). Whatever the cause of the heterozygote advantage, if s_1 and s_2 are the coefficients of selection against the AA and aa genotypes respectively, then an equilibrium is established (without recurrent mutation) when the ratio of the gene frequencies of A and a, p/q, is s_2/s_1. Selection in favour of heterozygotes for cystic fibrosis. For CF, the disease frequency in the UK is about one in 2000 births.

It is worth remembering that the medically important mendelian diseases are those that are both common and serious. They must all have some or other special trick to remain common in the face of selection. The trick may be an exceptionally high mutation rate (Duchenne muscular dystrophy), or propagation of non-pathological premutations (Fragile X), or onset of symptoms after reproductive age (Huntington disease) - but for common serious recessive conditions it is most often heterozygote advantage.

Nonmendelian Characters

Research into Simple and Complex Traits has Long Defined two Separate Traditions Within Human Genetics

By the time Mendel's work was rediscovered in 1900, a rival school of genetics was well established in the UK and elsewhere. Francis Galton, the remarkable and eccentric cousin of Charles Darwin, devoted much of his vast talent to systematising the study of human variation. Starting with an article on *Hereditary Talent and Character* published the same year, 1865, as Mendel's paper (and expanded in 1869 to a book, *Hereditary Genius*), he spent many years investigating family resemblances.

Galton was devoted to quantifying observations and applying statistical analysis. His Anthropometric Laboratory, established in 1884, recorded from his subjects (who paid him three pence for the privilege) their weight, sitting and standing height, arm span, breathing capacity, strength of pull and of squeeze, force of blow, reaction time, keenness of sight and hearing, colour discrimination and judgements of length. Except for colour blindness, these are quantitative, continuously variable characters. In one of the first applications of statistics he compared physical attributes of parents and children, and established the degree of correlation between relatives. By 1900 he had established a large body of knowledge about the inheritance of such attributes, and a tradition (biometrics) of their investigation.

A Historical Controversy

When Mendel's work was rediscovered, a controversy arose. The claims of the mendelians, championed by Bateson, were resisted by biometricians. Biometricians allowed that mendelian genes might explain a few rare abnormalities or curious quirks, but pointed out that most of the characters likely to be important in evolution (body size, build, strength, skill in catching prey or finding food) were continuous or quantitative characters and not amenable to mendelian analysis. You cannot define their inheritance by drawing pedigrees and marking in the affected people, because we all have these characters, only to different degrees. Mendelian analysis requires dichotomous characters (characters like extra fingers, that you either have or don't have). The controversy ran on, heatedly at times, until 1918. That year saw a seminal paper by RA Fisher demonstrating that continuous characters governed by a large number of independent mendelian factors (polygenic characters) would display precisely the quantitative variation and family correlations described by the biometricians. Later Falconer extended this model to cover dichotomous characters.

Two Traditions in Human Genetics

In principle Fisher's description of polygenic inheritance unified genetics. This was indeed generally true for the genetics of experimental organisms or farm animals. In human genetics, however, studies of mendelian and quantitative characters tended to continue as separate traditions, and until very recently few investigators felt at home in both worlds. The spectacular advances of 1970-1990 were entirely in mendelian genetics, whilst investigation of nonmendelian characters remained largely limited to statistical studies of family resemblances. Geneticists from the mendelian tradition were often reluctant to get involved in these studies, partly because of the complex statistical methodology and no doubt also because of a feeling that they were a poor investment of research effort compared to mapping and cloning genes for mendelian characters. Also many studies concerned sensitive areas of behavioural genetics such as the heritability of IQ, where violent controversies and a distastefully confrontational style of argument often reigned.

Multifactorial Nonmendelian Characters can be Oligogenic or Polygenic

The further away a character is from the primary gene action, the less likely is it to show a simple mendelian pedigree pattern. DNA sequence variants are virtually always cleanly mendelian - which is their major attraction as genetic markers. Protein variants (electrophoretic mobility or enzyme activity) are usually mendelian but can depend on more than one locus because of post-translational modification. The failure or malfunction of a developmental pathway that results in a birth defect is likely to involve a complex balance of factors. Thus the common birth defects (cleft palate, congenital dislocation of the hip, congenital heart disease, etc.) are rarely mendelian. Behavioural traits like IQ test performance or schizophrenia are still less likely to be mendelian. This does not however mean that they may not be genetic, either partly or entirely. Nonmendelian characters may depend on two, three or many genetic loci, with greater or smaller contributions from environmental factors. We use multifactorial here as a catch-all term covering all these possibilities. More specifically, the genetic determination may involve a small number of loci (oligogenic) or many loci each of individually small effect (polygenic); or there may be a single major locus with a multifactorial background. For dichotomous characters the underlying loci are envisaged as susceptibility genes, while for quantitative characters they are seen as quantitative trait loci (QTLs).

The New Synthesis Uses Mendelian Markers to Analyse Nonmendelian Phenotypes

Recent developments have finally brought together the study of mendelian and complex human phenotypes. Automation is allowing genetic analysis and sequencing on a scale scarcely imagined ten years ago. This has had two consequences. Most human genes are now identified, at least as expressed sequence tags (ESTs), so that molecular geneticists are looking for fresh fields to conquer. At the same time, marker studies can now be done on a scale that is probably large enough to deliver the statistical power needed to detect individual quantitative trait loci and susceptibility loci. Given the overwhelming preponderance of nonmendelian conditions in human disease, molecular dissection of complex phenotypes is widely seen as the next frontier in medical genetics. There are not two sorts of genes, mendelian genes and polygenes. It is possible however that there are two sorts of mutations. The sequence variants that confer susceptibility to polygenic disease may not, for the most part, be the gross mutations seen in mendelian conditions, which typically totally inactivate a gene. They are likely to include more subtle variants that slightly modify the expression of a gene. These will exist as common nonpathogenic variants in the healthy population, that cause problems only when combined with a whole series of similar variants at other loci.

Counselling in Nonmendelian Conditions uses Empiric Risks

In genetic counselling for nonmendelian conditions, risks are not derived from polygenic theory; they are empiric risks obtained through population surveys. This is fundamentally different from mendelian conditions, where the risks come from theory. The effect of family history is also quite different. If a couple are both carriers of cystic fibrosis, the risk of their next child being affected is 1 in 4. This remains true regardless how many affected or normal children they have already produced. If they have had a baby with neural tube defect, the recurrence risk is about 1 in 25 in the UK - but if they have already had two affected babies, the recurrence risk is about 1 in 12. It is not that having a second affected baby has caused their recurrence risk to increase, but it has enabled us to recognize them as a couple who always had been at particularly high risk. A cynic would say it involves the counsellor being wise after the event, but the practice accords with our understanding based on threshold theory, as well as with epidemiological data, and it is the best we can offer in an imperfect state of knowledge.

5

Cells and Viruses

Cellular Organization

Cells are the smallest structural unit of living organisms, capable of maintaining life and reproducing. Viruses are not cells because they cannot maintain life and reproduce by themselves. Although a nerve cell looks entirely different from a red blood cell, their organizations are essentially the same. Even plant cells and animal cells share significant similarity in the overall organization. All cells are divided into two types: prokaryotic cells and eukaryotic cells. The prokaryotic cell does not have a nucleus. The eukaryotic cell contains a nucleus. Eukaryotes are the organisms made up of eukaryotic cells. They include protista, fungi, animals and plants. Prokaryotes include archaebacteria and eubacteria. They are single-cell organisms.

More recently, "archaebacteria" have been placed in a category outside "bacteria", because they are quite different from the ordinary bacteria. According to the new classification, prokaryotes are divided into archaea and bacteria, where "archaea" is equivalent to "archaebacteria", and "bacteria" is the same as "eubacteria". Archaea live in extreme environments. They may be organized into three groups. Methanogens live in anaerobic environment such as swamps. They produce methane and cannot tolerate exposure to oxygen. Extreme halophiles live in very high concentrations of salt (NaCl), e.g., the Dead Sea and the Great Salt Lake. Extreme thermophiles live in hot, sulfur rich and low pH environment, such as hot springs, geysers and fumaroles in the Yellowstone National Park.

Basic Cellular Components

All cells contain cytoplasm, plasma membrane, and DNA. Cytoplasm is the viscous contents of a cell, including proteins, ribosomes, metabolites and ions. Ribosomes are the sites of protein synthesis. Plasma membrane is the cell membrane surrounding cytoplasm. It consists of phospholipid bilayer, associated proteins and carbohydrates. Phospholipid bilayer is also the basic constituent of other biomembranes. DNA (deoxyribonucleic acid) is the genetic material. An eukaryotic cell contains several DNA molecules, located in the nucleus and mitochondria which are membrane-bound organelles. A prokaryotic cell contains a single DNA molecule, which has no specific boundary with the cytoplasm.

Prokaryotic Cells

A prokaryotic cell consists of DNA, cytoplasm, and a surface structure which includes the plasma membrane and some of the following components:

- Cell wall
- Capsule
- Slime
- Flagella
- Fimbriae/Pili.

In bacteria, the cell wall contains a unique structure called peptidoglycan. Archaea do not possess peptidoglycan, but some archaea may contain pseudopeptidoglycan, which is composed of N-acetyltalosaminuronic acid, instead of N-acetylmuramic in peptidoglycan. Because of this structural difference, archaea are resistant to many cell wall antibiotics.

Bacteria may be divided into two groups, on the basis of their cell wall structures and the response to Gram stain: Gram-negative and Gram-positive. In Gram-negative bacteria, the cell wall is composed of three layers: (1) the periplasmic space which is an open area located outside the plasma membrane, (2) a thin layer of peptidoglycan external to the periplasmic space, and (3) the outer membrane surrounding the peptidoglycan. The Gram-positive bacteria do not have the periplasmic space and outer membrane, but have a thicker peptidoglycan layer. As a result, they are quite sensitive to lysozyme and penicillin. Capsule and slime are the hydrophilic gel surrounding the cell wall in most bacteria. The capsule is more closely associated with the cell than the slime. Flagella are long, rigid protein rods, facilitating the movement of motile bacteria. Fimbriae and pili are short hair-like structures used to attach other cells. They are essential for infecting other organisms.

Spores

The spore is a small, often unicellular, reproductive unit of plants, algae, fungi, protozoa, and bacteria. Bacterial spores have thick walls which can withstand varying temperatures, humidity, and other unfavourable conditions.

Eukaryotic Cells

The eukaryotic cell contains organelles, which are defined as membrane-bound structures such as nucleus, mitochondria, chloroplasts, endoplasmic reticulum (ER), Golgi apparatus, lysosomes, vacuoles, peroxisomes, etc. Prokaryotic cells do not have organelles.

For animal cells, the cell surface consists of the plasma membrane only, but plant cells have an additional layer called cell wall, which is made up of cellulose and other polymers. The nucleus is the largest organelle in an eukaryotic cell. It is not part of the cytoplasm. By definition, cytoplasm is everything inside the plasma membrane except the nucleus. Under microscope, the nucleus shows two distinct areas. The darker area is called nucleolus, and the lighter area is known as nucleoplasm. Cytosol is the cytoplasm excluding organelles. It contains cytoskeleton, ribosomes, proteins and other smaller molecules.

Biomembranes

All biological membranes, including plasma membranes and all organelle membranes, contain lipids and proteins. The lipids found in biomembranes are mainly phospholipids and cholesterol. In the plasma membrane and some of organelle membranes, proteins and phospholipids are attached to carbohydrates, forming glycoproteins and glycolipids, respectively.

Phospholipids

A phospholipid molecule consists of a hydrophilic polar head group and a hydrophobic tail. The polar head group contains one or more phosphate groups. The hydrophobic tail is made up of two fatty acyl chains. When many phospholipid molecules are placed in water, their hydrophilic heads tend to face water and the hydrophobic tails are forced to stick together, forming a bilayer.

Polar Head Groups

Most phospholipid head groups belong to phosphoglycerides, which contain glycerol joining the head and the tail. Examples of phosphoglycerides include phosphatidylcholine, phosphatidylserine, phosphatidylethanolamine, phosphatidylinositol, etc.

Fatty Acyl Chains

The fatty acyl chain in biomembranes usually contains even number of carbon atoms. They may be saturated (neighbouring C atoms are all connected by single bonds) or unsaturated (some neighbouring C atoms are connected by double bonds).

***Table** : Cellular fatty acids.*

Chemical Formula	*Name*
Saturated fatty acid	
$CH_3(CH_2)_{10}COOH$	Lauric
$CH_3(CH_2)_{12}COOH$	Myristic
$CH_3(CH_2)_{14}COOH$	Palmitic
$CH_3(CH_2)_{16}COOH$	Stearic
$CH_3(CH_2)_{18}COOH$	Arachidic
$CH_3(CH_2)_{22}COOH$	Lignoceric
Unsaturated fatty acid	
$CH_3(CH_2)_5CH{=}CH(CH_2)_7COOH$	Palmitoleic
$CH_3(CH_2)_7CH{=}CH(CH_2)_7COOH$	Oleic
$CH_3(CH_2)_4CH{=}CHCH_2CH{=}CH(CH_2)_7COOH$	Linoleic
$CH_3(CH_2)_4(CH{=}CHCH_2)_3CH{=}CH(CH_2)_3COOH$	Arachidonic
$CH_3CH_2CH{=}CHCH_2CH{=}CHCH_2CH{=}CH(CH_2)_7COOH$	Linolenic

Cholesterol and Steroids

Cholesterol is absent from most prokaryotic cells, but abundant in the plasma membrane of mammalian cells. It is used as a precursor to generate other important steroids. Cholesterol plays a central role in atherosclerosis-a disorder that may cause heart attack or stroke. It is also involved in Alzheimer's disease.

Glycoproteins and Glycolipids

Glycoproteins are the proteins covalently attached to carbohydrates such as glucose, galactose, lactose, fucose, sialic acid, N-acetylglucosamine, N-acetylgalactosamine, etc. Glycolipids are carbohydrate-attached lipids. Their role is to provide energy and also serve as markers for cellular recognition. The antigens which determine blood types belong to glycoproteins and glycolipids (more info).

Blood Group Antigens

The antigens which determine blood types belong to glycoproteins and glycolipids. There are three types of blood-group antigens: O, A, and B. They differ only slightly in the composition of carbohydrates.

The Nucleus

The cell nucleus consists of nuclear envelope, nucleolus and nucleoplasm. Most chromosomes are located in the nucleoplasm, but portions of several chromosomes containing clusters of rRNA genes may get together in the nucleolus, forming the nucleolar organizing region. The major role of the nucleolus is to produce rRNA.

Introduction to Chromosomes

Chromosomes are the structures that hold DNA molecules. One chromosome contains a DNA molecule. Each chromosome has a p and q arm; p is the shorter arm and q is the longer arm. The arms are separated by a pinched region called the centromere. In order for chromosomes to be seen with a microscope, they need to be stained. Once stained, the chromosomes look like strings with light and dark "bands" and their picture can be taken. The picture, or chromosome map, is called a karyotype.

The germ cell (sperm or egg) of a human being contains 23 chromosomes, labeled from 1 to 22 and either X or Y. The somatic cells (cells other than germ cells) of a normal person has 46 chromosomes. For other species, the chromosome number varies from 1 to 1260. A human somatic cell contains two chromosomes that determine the sex of a person. The two sex chromosomes are XY in male and XX in female. The gene (SRY) that is important for testis formation is located on chromosome Y. It is possible that a person with testis still exhibits female characteristics.

Chromosomes and Karyotype

In a non-dividing cell, chromosomes are not visible by light microscopy, because chromatin spreads throughout the nucleus. During the metaphase of cell division, the chromatin condenses and becomes visible as chromosomes. At this time, each chromosome has been duplicated. A chromosome becomes two sister chromatids attached at the centromere.

Chromosome Banding

To see chromosomes by microscope, they are normally treated with chemical dyes, such as Giemsa. The chromosome will appear as a series of alternate dark and light bands. If Giemsa is used, the dark band is called G-band or G-positive band, and the light band is named G-negative band. Similar banding patterns can be observed by using another dye, Quinacrine.

However, if chromosomes were treated in a hot alkaline solution before staining with Giemsa, a reverse pattern will be observed, namely, the original dark band will become light band, and vice versa. For this reason, the G-negative band is also known as the R-band. Chromosome bands are named as follows. Each chromosome consists of two arms separated by the centromere.

The long arm and short arm are labeled q (for queue) and p (for petit), respectively. At the lowest resolution, only a few major bands can be distinguished, which are labeled q1, q2, q3; p1, p2, p3, etc., counting from the centromere.

Higher resolution reveals sub-bands, labeled q11, q12, q13, etc. Sub-sub-bands identified by even higher resolution are labeled q11.1, q11.2, q11.3, etc. Traditionally, the short arm (p) is displayed on top of the long arm (q).

Karyotype

Karyotype is the representation of entire metaphase chromosomes in a cell, arranged in order of size.

Chromosome Numbers

A germ cell (sperm or egg) contains only one set of chromosomes. It belongs to haploid, represented as 1n. Somatic cells (cells other than germ cells) of sexually reproducing organisms are diploid, denoted by 2n. In humans, the haploid chromosome number is 23, but the diploid chromosome number is 46.

Extremes

Smallest number: The female of a subspecies of the ant, *Myrmecia pilosula*, has one pair of chromosomes per cell. Its male has only one chromosome in each cell. Largest number: In the fern family of plants, the species *Ophioglossum reticulatum* has about 630 pairs of chromosomes, or 1260 chromosomes per cell.

Sex Chromosomes

Sex chromosomes determine the sex of an organism. A human somatic cell has two sex chromosomes: XY in male and XX in female. A human germ cell has one sex chromosome: X or Y in a sperm and X in an egg. When an X-sperm is combined with an egg, the resulting zygote (fertilized egg) will contain two X chromosomes. A person developed from the XX-zygote will have the characteristics of a female. Combination of a Y-sperm and an egg will produce a male.

The SRY Gene

Usually, a woman has two X chromosomes (XX) and a man one X and one Y (XY). However, both male and female characteristics can sometimes be found in one individual, and it is possible to have XY women and XX men. Analysis of such individuals has revealed some of the molecules involved in sex determination, including one called SRY, which is important for testis formation. SRY (which stands for sex-determining region Y gene) is found on the Y chromosome. In the cell, it binds to other DNA and in doing so distorts it dramatically out of shape. This alters the properties of the DNA and likely alters the expression of a number of genes, leading to testis formation. Most XX men who lack a Y chromosome do still have a copy of the SRY gene on one of their X chromosomes (moved there by chromosomal translocation).

This copy accounts for their maleness. However, because the remainder of the Y chromosome is missing they frequently do not develop secondary sexual characteristics in the usual way.

Cytoplasmic Organelles

By definition, organelles are the membrane-bound structures in a cell. The nucleus is an example. Other organelles are located in the cytoplasm such as mitochondria, chloroplasts, endoplasmic reticulum, Golgi apparatus, peroxisomes, lysosomes, vacuoles and glyoxisomes.

Mitochondria

An eukaryotic cell contains many mitochondria, occupying up to a quarter of the cytoplasmic volume. The size of a mitochondrion is about 1.5-2 mm in length, 0.5-1 mm in diameter, approximately the same as *E. coli*. It has two membranes: outer membrane and inner membrane. Mitochondria also have their own DNA (represented as mtDNA), which encodes some of the proteins and RNAs in mitochondria. However, most proteins operating in mitochondria still originate from nuclear DNA. The major role of mitochondria is to produce ATP (adenosine triphosphate), which carries high energy to power most cellular processes. Such energy is stored in the phosphoanhydride bonds of ATP. During ATP hydrolysis, the bond is broken, releasing 7.3 kcal/mole of energy. Many cellular processes can utilize the released energy by coupling with the ATP hydrolysis. In animal cells, the major sources for the synthesis of ATP are fatty acids and glucose. Oxidation of an 18-carbon fatty acid can make 146 ATP molecules.

By contrast, oxidation of one glucose molecule (6 carbons) can generate only 36 ATP molecules. The generation of ATP involves a series of electron transport. Inevitably, electrons may leak from the electron transport chain, producing free radicals. This has been suggested to be the major mechanism involved in the aging process.

Chloroplasts

Like mitochondria, a chloroplast also contains both outer and inner membranes on its surface. Inside the chloroplast, there are many thylakoids, each is enclosed by a membrane. Chlorophylls are located on the thylakoid membrane to absorb light for photosynthesis. In the first step of photosynthesis, light energy is used to split water into hydrogen ions and oxygen molecules.

The generated hydrogen ions will create a concentration gradient across the thylakoid membrane. Movement of hydrogen ions through the membrane is coupled to ATP synthesis.

Endoplasmic Reticulum

Endoplasmic reticulum (ER) can be divided into rough ER and smooth ER. The major role of rough ER is to process the newly synthesized peptides from ribosomes.

Therefore, the surface of rough ER is usually associated with ribosomes and thus appears "rough". Smooth ER is involved in the synthesis and metabolism of lipids. Hepatocytes are abundant in smooth ER.

Golgi Apparatus

Golgi apparatus is a major site for sorting and modifications of proteins and lipids. After proteins are sorted at rough ER, they are enclosed in transport vesicles and carried to the Golgi apparatus. Some proteins could be modified into glycoproteins and then transported to other destinations.

Peroxisomes

Peroxisomes contain enzymes for degrading amino acids and fatty acids. These reactions produce harmful hydrogen peroxide. Hence, peroxisomes also contain catalase to convert hydrogen peroxide into water and oxygen:

Lysosomes

The major function of lysosomes is to degrade various macromolecules in the cell. They contain nuclease for degrading DNA and RNA, protease for degrading proteins and other enzymes for degrading polysacchrides and lipids. Lysosomes exist only in animal cells. Although plant cells do not have lysosomes, their vacuoles are also capable of degrading macromolecules..

Vacuoles

Vacuoles store small molecules such as water, ions, sucrose and amino acids. They can also hold waste products which will be slowly degraded. They typically occupy more than 30% of the cell volume, but may expand up to 90%.

Glyoxisomes

Glyoxisomes are found mainly in plant seeds. Their major function is to convert fatty acids into acetyl CoA for the glyoxylate cycle where two acetyl-CoA molecules are converted to a 4-carbon dicarboxylic acid. Peroxisomes and glyoxisomes are also called microbodies.

Viruses

Viruses are the smallest organisms, with diameters ranging from 20 nm to 300 nm (1 nm = 10^{-9} meter). Viruses are not cells. They consist of one or more molecules of DNA or RNA, which contain the virus's genes, surrounded by a protein coat called capsid. Some viruses also have an envelope surrounding the capsid. Viruses can be sphere-shaped, or helical. Unlike most bacteria, most viruses do cause disease because they invade living, normal cells, such as those in the human body. They then multiply and produce other viruses like themselves. Each virus is very particular about which cell it attacks. Various human viruses specifically attack particular cells in the body's organs, systems, or tissues, such as the liver, respiratory system, or blood cells. Although types of viruses behave differently, most survive by taking over the machinery that makes a cell work. Briefly, when a single virus particle, a "virion", comes in contact with a cell it likes, it may attach to special landing sites on the surface of that cell. From there, the virus may inject molecules into the cell, or the cell may swallow up the virion. Once inside the cell, viral molecules such as DNA or RNA direct the cell to make new virus offspring. That's how a virus "infects" a cell. Viruses can even "infect" bacteria. These viruses, called bacteriophages, may help researchers develop alternatives to antibiotic medicines for wiping out bacterial infections.

Many viral infections do not result in disease. For example, by the time most people in the United States become adults, they have been infected by cytomegalovirus (CMV). Most of these people, however, do not develop CMV disease symptoms. Other viral infections can result in deadly diseases, such as HIV which causes acquired immunodeficiency syndrome (AIDS) and coronaviruses which cause severe acute respiratory syndrome (SARS).

Baltimore Classification of Viruses

The Baltimore classification is based on genetic contents and replication strategies of viruses. The genetic material in all types of cells is double-stranded DNA, but some viruses use RNA or single-stranded DNA to carry genetic information.

According to Baltimore classification, viruses are divided into the following seven classes:

1. dsDNA viruses
2. ssDNA viruses
3. dsRNA viruses
4. (+)-sense ssRNA viruses
5. (-)-sense ssRNA viruses
6. RNA reverse transcribing viruses
7. DNA reverse transcribing viruses.

Where "ds" represents "double strand" and "ss" denotes "single strand".

The Life Cycle of Viruses

The life cycle of viruses may be divided into the following stages:

Attachment

Attachment is a specific binding between viral surface proteins and their receptors on the host cellular surface. This specificity determines the host range of a virus. For instance, the human immunodeficiency virus (HIV) attacks only human's immune cells (mainly T cells), because its surface protein, gp120, can interact with CD4 and chemokine receptors on the T cell's surface.

Penetration

Following attachment, viruses may enter the host cell through receptor mediated endocytosis or other mechanisms.

Uncoating

Uncoating is a process that viral capsid is degraded by viral enzymes or host enzymes.

Replication

Replication involves assembly of viral proteins and genetic materials produced in the host cell.

Release

Viruses may escape from the host cell by causing cell rupture (lysis). Enveloped viruses (e.g., HIV) typically "bud" from the host cell. During the budding process, a virus acquires the phospholipid envelope containing the embedded viral glycoproteins.

6

Occurrence and Evolution of Cell

Peroxisomes are found in virtually all eukaryotic cells. Peroxisomes contain enzymes for certain oxidative reactions, like the beta-oxidation of very-long-chain fatty acids. Prokaryotes lack peroxisomes. The enzymatic content of peroxisomes varies across species, but the presence of certain proteins common to many species has been used to suggest an endosymbiotic origin; that is, peroxisomes evolved from bacteria that invaded larger cells as parasites, and very gradually evolved a symbiotic relationship. However, this view has been challenged by recent discoveries. For example, peroxisome-less mutants can restore peroxisomes upon introduction of the wild-type gene.

Two independent evolutionary analyses of the peroxisomal proteome found homologies between the peroxisomal import machinery and the ERAD pathway in the endoplasmic reticulum, along with a number of metabolic enzymes that were likely recruited from the mitochondria. These results indicate that the peroxisome does not have an endosymbiotic origin; instead, it likely originates from the ER, and its proteins were recruited from pools existing within the primitive eukaryote, as quoted in the science textbook Biozone. However, latest research have suggested that the peroxisome may have had an actinobacterial origin, although this is still controversial. Other organelles of the Microbody family related to peroxisomes include glyoxysomes of plants and filamentous fungi, glycosomes of kinetoplastids and Woronin bodies of filamentous fungi.

Peroxisomes contain oxidative enzymes, such as catalase, D-amino acid oxidase, and uric acid oxidase. However the last enzyme is absent in humans, explaining the disease known as gout, caused by the accumulation of uric acid.

Certain enzymes within the peroxisome, by using molecular oxygen, remove hydrogen atoms from specific organic substrates (labeled as R), in an oxidative reaction, producing hydrogen peroxide (H_2O_2, itself toxic):

$$RH_2 + O_2 \rightarrow R + H_2O_2$$

Catalase, another enzyme in the peroxisome, in turn uses this H_2O_2 to oxidize other substrates, including phenols, formic acid, formaldehyde, and alcohol, by means of the peroxidation reaction:

$$H_2O_2 + R'H_2 \rightarrow R' + 2H_2O,$$

thus eliminating the poisonous hydrogen peroxide in the process.

This reaction is important in liver and kidney cells, where the peroxisomes detoxify various toxic substances that enter the blood. About 25% of the ethanol we drink is oxidized to acetaldehyde in this way. In addition, when excess H_2O_2 accumulates in the cell, catalase converts it to H_2O through this reaction:

$$2H_2O_2 \rightarrow 2H_2O + O_2$$

A major function of the peroxisome is the breakdown of fatty acid molecules, in a process called beta-oxidation. In this process, the fatty acids are broken down two carbons at a time, converted to Acetyl-CoA, which is then transported back to the cytosol for further use. In animal cells, beta-oxidation can also occur in the mitochondria. In yeast and plant cells, this process is exclusive for the peroxisome. The first reactions in the formation of plasmalogen in animal cells also occur in peroxisomes. Plasmalogen is the most abundant phospholipid in myelin. Deficiency of plasmalogens causes profound abnormalities in the myelination of nerve cells, which is one of the reasons that many peroxisomal disorders lead to neurological disease (adrenoleukodystrophy).

Peroxisomes also play a role in the production of bile acids and proteins. In higher plants, peroxisomes contain also a complex battery of antioxidative enzymes such as superoxide dismutase, the components of the ascorbate-glutathione cycle, and the NADP-dehydrogenases of the pentose-phosphate pathway. It has been demonstrated the generation of superoxide (O_2) and nitric oxide ($^{\cdot}NO$) radicals. The peroxisome of plant cells is polarised when fighting fungal penetration. Infection causes a glucosinolate molecule to play an antifungal role to be made and delivered to the outside of the cell through the action of the peroxisomal proteins (PEN2 and PEN3).

Protein Import

Proteins are selectively imported into peroxisomes. Since the organelles contain no DNA or ribosomes and, thus, have no means of producing proteins, all of their proteins must be imported across the membrane. It is believed that necessary proteins enter through the endoplasmic reticulum during biogenesis as well as through membrane proteins.

A specific protein signal (PTS or peroxisomal targeting signal) of three amino acids at the *C-terminus* of many peroxisomal proteins signals the membrane of the peroxisome to import them into the organelle. Other peroxisomal proteins contain a signal at the *N-terminus*. There are at least 32 known peroxisomal proteins, called peroxins, which participate in the process of importing proteins by means of *ATP hydrolysis*. Proteins do not have to unfold to be imported into the peroxisome. The protein receptors, the peroxins *PEX5* and *PEX7*, accompany their cargoes (containing a PTS1 or a PTS2, respectively) all the way into the peroxisome where they release the cargo and then return to the cytosol – a step named *recycling*. Overall, the import cycle is referred to as the *extended shuttle mechanism*. Evidence now indicates that ATP hydrolysis is required for the recycling of receptors to the cytosol. Also, ubiquitination appears to be crucial for the export of PEX5 from the peroxisome, to the cytosol. Little is known about the import of PEX7, although it has helper proteins that have been shown to be ubiquitinated. It is interesting to note that PEX7 helper proteins share many similarities with PEX5, and, in higher eukaryotes, they are functionally replaced by a long isoform of PEX5 (PEX5L).

Deficiencies

Peroxisomal disorders are a class of conditions that lead to disorders of lipid metabolism and diseases of the nervous system. Well-known examples are X-linked adrenoleukodystrophy, the most frequent, and Zellweger syndrome. Peroxisomes matrix proteins are synthesized on free ribosomes in the cytosol and are imported post-translationally in pre-existing vesicles.

Genes

Genes that encode peroxisomal proteins include:

- PEX1
- PEX2 - PXMP3
- PEX3
- PEX5
- PEX6
- PEX7
- PEX10
- PEX11A, PEX11B, PEX11G
- PEX12
- PEX13

- PEX14
- PEX16
- PEX19
- PEX26
- PEX28
- PEX30
- PEX31

Glyoxysome

Glyoxysomes are specialized peroxisomes found in plants (particularly in the fat storage tissues of germinating seeds) and also in filamentous fungi. As in all peroxisomes, in glyoxysomes the fatty acids are hydrolyzed to acetyl-CoA by peroxisomal α-oxidation enzymes. Besides peroxisomal functions, glyoxysomes possess additionally the key enzymes of glyoxylate cycle (isocitrate lyase and malate synthase) which accomplish the glyoxylate cycle bypass. Thus, glyoxysomes (as all peroxisomes) contain enzymes that initiate the breakdown of fatty acids and additionally possess the enzymes to produce intermediate products for the synthesis of sugars by gluconeogenesis. The seedling uses these sugars synthesized from fats until it is mature enough to produce them by photosynthesis.

Ribosomes

Ribosomes are the components of cells that make proteins from all amino acids. One of the central tenets of biology, often referred to as the "central dogma," is that DNA is used to make RNA, which, in turn, is used to make protein. The DNA sequence in genes is copied into a messenger RNA (mRNA). Ribosomes then read the information in this RNA and use it to create proteins. This process is known as translation; i.e., the ribosome "translates" the genetic information from RNA into proteins. Ribosomes do this by binding to an mRNA and using it as a template for the correct sequence of amino acids in a particular protein. The amino acids are attached to transfer RNA (tRNA) molecules, which enter one part of the ribosome and bind to the messenger RNA sequence. The attached amino acids are then joined together by another part of the ribosome. The ribosome moves along the mRNA, "reading" its sequence and producing a chain of amino acids.

Ribosomes are made from complexes of RNAs and proteins. Ribosomes are divided into two subunits, one larger than the other. The smaller subunit binds to the mRNA, while the larger subunit binds to the tRNA and the amino acids. When a ribosome finishes reading a mRNA, these two subunits split apart. Ribosomes have been classified as ribozymes, since the ribosomal RNA seems to be most important for the peptidyl transferase activity that links amino acids together.

Ribosomes from bacteria, archaea and eukaryotes (the three domains of life on Earth), have significantly different structures and RNA sequences. These differences in structure allow some antibiotics to kill bacteria by inhibiting their ribosomes, while leaving human ribosomes unaffected. The ribosomes in the mitochondria of eukaryotic cells resemble those in bacteria, reflecting the likely evolutionary origin of this organelle. The word ribosome comes from *ribo*nucleic acid and the Greek: *soma* (meaning body).

Description

Archaeal, eubacterial and eukaryotic ribosomes differ in their size, composition and the ratio of protein to RNA. Because they are formed from two subunits of non-equal size, they are slightly longer in the axis than in diameter. Prokaryotic ribosomes are around 20 nm (200 ångströms) in diameter and are composed of 65% ribosomal RNA and 35% ribosomal proteins (known as a ribonucleoprotein or RNP). Eukaryotic ribosomes are between 25 and 30 nm (250-300 ångströms) in diameter and the ratio of rRNA to protein is close to 1. Ribosomes translate messenger RNA (mRNA) and build polypeptide chains (e.g., proteins) using amino acids delivered by transfer RNA (tRNA). Their active sites are made of RNA, so ribosomes are now classified as "ribozymes".

Ribosomes build proteins from the genetic instructions held within messenger RNA. Free ribosomes are suspended in the cytosol (the semi-fluid portion of the cytoplasm); others are bound to the rough endoplasmic reticulum, giving it the appearance of roughness and thus its name, or to the nuclear envelope. As ribozymes are partly constituted from RNA, it is thought that they might be remnants of the RNA world. Although catalysis of the peptide bond involves the C2 hydroxyl of RNA's P-site adenosine in a protein shuttle mechanism, other steps in protein synthesis (such as translocation) are caused by changes in protein conformations.

Ribosomes are sometimes referred to as organelles, but the use of the term *organelle* is often restricted to describing sub-cellular components that include a phospholipid membrane, which ribosomes, being entirely particulate, do not. For this reason, ribosomes may sometimes be described as "non-membranous organelles".

Ribosomes were first observed in the mid-1950s by Romanian cell biologist George Palade using an electron microscope as dense particles or granules for which he would win the Nobel Prize. The term "ribosome" was proposed by scientist Richard B. Roberts in 1958:

During the course of the symposium a semantic difficulty became apparent. To some of the participants, "microsomes" mean the ribonucleoprotein particles of the microsome fraction contaminated by other protein and lipid material; to others, the microsomes consist of protein and lipid contaminated by particles. The phrase "microsomal particles" does not seem adequate, and "ribonucleoprotein particles of the microsome fraction" is much too awkward. During the meeting, the word "ribosome" was suggested, which has a very satisfactory name and a pleasant sound. The present confusion would be eliminated if "ribosome" were adopted to designate ribonucleoprotein particles in sizes ranging from 35 to 100S.

– Roberts, R. B., *Microsomal Particles and Protein Synthesis*

The structure and function of the ribosomes and associated molecules, known as the *translational apparatus*, has been of research interest since the mid-twentieth century and is a very active field of study today.

Ribosomes consist of two subunits that fit together and work as one to translate the mRNA into a polypeptide chain during protein synthesis. Bacterial subunits consist of one or two and eukaryotic of one or three very large RNA molecules (known as ribosomal RNA or rRNA) and multiple smaller protein molecules. Crystallographic work has shown that there are no ribosomal proteins close to the reaction site for polypeptide synthesis. This suggests that the protein components of ribosomes act as a scaffold that may enhance the ability of rRNA to synthesize protein rather than directly participating in catalysis.

Biogenesis

In bacterial cells, ribosomes are synthesized in the cytoplasm through the transcription of multiple ribosome gene operons. In eukaryotes, the process takes place both in the cell cytoplasm and in the nucleolus, which is a region within the cell nucleus. The assembly process involves the coordinated function of over 200 proteins in the synthesis and processing of the four rRNAs, as well as assembly of those rRNAs with the ribosomal proteins.

Ribosome Locations

Ribosomes are classified as being either "free" or "membrane-bound". Free and membrane-bound ribosomes differ only in their spatial distribution; they are identical in structure. Whether the ribosome exists in a free or membrane-bound state depends on the presence of an ER-targeting signal sequence on the protein being synthesized, so an individual ribosome might be membrane-bound when it is making one protein, but free in the cytosol when it makes another protein.

Free Ribosomes

Free ribosomes can move about anywhere in the cytosol, but are excluded from the cell nucleus and other organelles. Proteins that are formed from free ribosomes are released into the cytosol and used within the cell. Since the cytosol contains high concentrations of glutathione and is, therefore, a reducing environment, proteins containing disulfide bonds, which are formed from oxidized cysteine residues, cannot be produced in this compartment.

Membrane-bound Ribosomes

When a ribosome begins to synthesize proteins that are needed in some organelles, the ribosome making this protein can become "membrane-bound". In eukaryotic cells this happens in a region of the endoplasmic reticulum (ER) called the "rough ER". The newly produced polypeptide chains are inserted directly into the ER by the ribosome and are then transported to their destinations, through the secretory pathway. Bound ribosomes usually produce proteins that are used within the plasma membrane or are expelled from the cell via *exocytosis*.

Structure

The ribosomal subunits of prokaryotes and eukaryotes are quite similar. The unit of measurement is the Svedberg unit, a measure of the rate of sedimentation in centrifugation rather than size and accounts for why fragment names do not add up (70S is made of 50S and 30S). Prokaryotes have 70S ribosomes, each consisting of a small (30S) and a large (50S) subunit. Their large subunit is composed of a 5S RNA subunit (consisting of 120 nucleotides), a 23S RNA subunit (2900 nucleotides) and 34 proteins. The 30S subunit has a 1540 nucleotide RNA subunit (16S) bound to 21 proteins.

Eukaryotes have 80S ribosomes, each consisting of a small (40S) and large (60S) subunit. Their large subunit is composed of a 5S RNA (120 nucleotides), a 28S RNA (4700 nucleotides), a 5.8S subunit (160 nucleotides) and ~49 proteins. The 40S subunit has a 1900 nucleotide (18S) RNA and ~33 proteins. The ribosomes found in chloroplasts and mitochondria of eukaryotes also consist of large and small subunits bound together with proteins into one 70S particle. These organelles are believed to be descendants of bacteria and as such their ribosomes are similar to those of bacteria.

The various ribosomes share a core structure, which is quite similar despite the large differences in size. Much of the RNA is highly organized into various tertiary structural motifs, for example pseudoknots that exhibit coaxial stacking. The extra RNA in the larger ribosomes is in several long continuous insertions, such that they form loops out of the core structure without disrupting or changing it. All of the catalytic activity of the ribosome is carried out by the RNA; the proteins reside on the surface and seem to stabilize the structure.

The differences between the bacterial and eukaryotic ribosomes are exploited by pharmaceutical chemists to create antibiotics that can destroy a bacterial infection without harming the cells of the infected person. Due to the differences in their structures, the bacterial 70S ribosomes are vulnerable to these antibiotics while the eukaryotic 80S ribosomes are not. Even though mitochondria possess ribosomes similar to the bacterial ones, mitochondria are not affected by these antibiotics because they are surrounded by a double membrane that does not easily admit these antibiotics into the organelle.

High-resolution Structure

The general molecular structure of the ribosome has been known since the early 1970s. In the early 2000s the structure has been achieved at high resolutions, on the order of a few ångströms. The first papers giving the structure of the ribosome at atomic resolution were published in rapid succession in late 2000. First, the 50S (large prokaryotic) subunit from the archaeon *Haloarcula marismortui* was published. Soon after, the structure of the 30S subunit from *Thermus thermophilus* was published. Shortly thereafter, a more detailed structure was published. These structural studies were awarded the Nobel Prize in Chemistry in 2009. Early the next year (May 2001) these coordinates were used to reconstruct the entire *T. thermophilus* 70S particle at 5.5 ångström resolution.

Two papers were published in November 2005 with structures of the *Escherichia coli* 70S ribosome. The structures of a vacant ribosome were determined at 3.5-angström resolution using x-ray crystallography. Then, two weeks later, a structure based on cryo-electron microscopy was published, which depicts the ribosome at 11-15 angström resolution in the act of passing a newly synthesized protein strand into the protein-conducting channel.

First atomic structures of the ribosome complexed with tRNA and mRNA molecules were solved by using X-ray crystallography by two groups independently, at 2.8 angstrom and at 3.7 angstrom. These structures allow one to see the details of interactions of the *Thermus thermophilus* ribosome with mRNA and with tRNAs bound at classical ribosomal sites. Interactions of the ribosome with long mRNAs containing Shine-Dalgarno sequences were visualized soon after that at 4.5- to 5.5-angström resolution.

Function

Ribosomes are the workhorses of protein biosynthesis, the process of translating mRNA into protein. The mRNA comprises a series of codons that dictate to the ribosome the sequence of the amino acids needed to make the protein. Using the mRNA as a template, the ribosome traverses each codon (3 nucleotides) of the mRNA, pairing it with the appropriate amino acid provided by a tRNA. Molecules of transfer RNA (tRNA) contain a complementary anticodon on one end and the appropriate amino acid on the other. The small ribosomal subunit, typically bound to a tRNA containing the amino acid methionine, binds to an AUG codon on the mRNA and recruits the large ribosomal subunit. The ribosome then contains three RNA binding sites, designated A, P and E. The A site binds an aminoacyl-tRNA (a tRNA bound to an amino acid); the P site binds a peptidyl-tRNA (a tRNA bound to the peptide being synthesized); and the E site binds a free tRNA before it exits the ribosome. Protein synthesis begins at a start codon AUG near the 5' end of the mRNA. mRNA binds to the P site of the ribosome first. The ribosome is able to identify the start codon by use of the Shine-Dalgarno sequence of the mRNA in prokaryotes and Kozak box in eukaryotes.

The ribosome uses tRNA that matches the current codon (triplet) on the mRNA to append an amino acid to the polypeptide chain. This is done for each triplet on the mRNA, while the ribosome moves towards the 3' end of the mRNA. Usually in bacterial cells, several ribosomes are working parallel on a single mRNA, forming what is called a *polyribosome* or *polysome.*

Cytoskeleton: Microtubules, Microfilaments and Microvilli

The cytoskeleton (also CSK) is a cellular "scaffolding" or "skeleton" contained within the cytoplasm and is made out of protein. The cytoskeleton is present in all cells; it was once thought to be unique to eukaryotes, but recent research has identified the prokaryotic cytoskeleton. It is a dynamic structure that maintains cell shape, protects the cell, enables cellular motion (using structures such as flagella, cilia and lamellipodia), and plays important roles in both intracellular transport (the movement of vesicles and organelles, for example) and cellular division. The concept of a protein mosaic that dynamically coordinated cytoplasmic biochemistry was proposed by Rudolph Peters in 1929 while the term (*cytosquelette,* in French) was first introduced by French embryologist Paul Wintrebert in 1931.

The Eukaryotic Cytoskeleton

Eukaryotic cells contain three main kinds of cytoskeletal filaments, which are microfilaments, intermediate filaments, and microtubules. The cytoskeleton provides the cell with structure and shape, and by excluding macromolecules from some of the cytosol it adds to the level of macromolecular crowding in this compartment. Cytoskeletal elements interact extensively and intimately with cellular membranes.

Microfilament

Microfilaments (or actin filaments) are the thinnest filaments of the cytoskeleton found in the cytoplasm of all eukaryotic cells. These linear polymers of actin subunits are flexible and relatively strong, resisting buckling by multipiconewton compressive forces and filament fracture by nanonewton tensile forces. Microfilaments are highly versatile, functioning in (a) cell crawling, amoeboid movement, and changes in cell shape, where one end of the actin filament elongates while the other end contractile, presumably by myosin II molecular motors (though an alternative "actoclampin"-driven expansile molecular motors exists), and (b) actomyosin-driven contractile molecular motors, where the *thin filaments* serve as tensile platforms for myosin's ATP hydrolysis-dependent pulling action in muscle contraction and uropod advancement.

Organization

Actin filaments are assembled in two general types of structures: bundles and networks. Bundles can be composed of polar (all barbed ends point to the same end of the bundle) or non-polar (barbed ends point towards both ends) arrays of filaments. A class of Actin-binding proteins, called cross-linking proteins, dictate the formation of these structures. Cross-linking proteins determine filament orientation and spacing in the bundles and networks. These structures are regulated by many other classes of actin-binding proteins (motor proteins, branching proteins, severing proteins, polymerization promoters, capping proteins, etc.).

In Vitro *Self-assembly*

The thinnest fibres of the cytoskeleton (measuring approximately 6 nm in diameter), microfilaments are formed by the head-to-tail polymerization of actin monomers (also known as globular or G-actin). Actin subunits as part of a fibre are referred to as filamentous actin (or F-actin). Each microfilament is made up of two helical interlaced strands of subunits. Much like microtubules, actin filaments are polarized, with their fast-growing barbed-ends (because of their appearance in electron micrographs after binding of myosin S1 sub-fragments) and a slow-growing pointed-end (again based on the pattern created by S1 binding). The pointed end is sometimes referred to as the minus (-) end and the barbed end is sometimes referred to as the plus (+) end because of the growth rates, but this is nomenclature adapted from the microtubule field, and is not generally accepted in the actin field.

In vitro actin polymerization, nucleation, starts with the self-association of three G-actin monomers to form a trimer. ATP-actin then binds the barbed end, and the ATP is subsequently hydrolyzed with a half time of about 2 seconds and the inorganic phosphate released with a half-time of about 6 minutes, which reduces the binding strength between neighbouring units and generally destabilizes the filament. *In vivo* actin polymerization is catalyzed by a new class of filament end-tracking molecular motors known as actoclampins. Recent evidence suggests that ATP hydrolysis can be prompt in such cases (i.e., the rate of monomer incorporation is matched by the rate of ATP hydrolysis).

ADP-actin dissociates slowly from the pointed end, but this process is greatly accelerated by ADP-cofilin, which severs ADP-rich regions nearest the (–)-ends. Upon release, ADP-actin undergoes exchange of its bound ADP for solution-phase ATP, thereby forming the ATP-actin monomeric units needed for further barbed-end filament elongation. This rapid turnover is important for the cell's movement. End-capping proteins such as CapZ prevent the addition or loss of monomers at the filament end where actin turnover is unfavourable like in the muscle apparatus.

Mechanism of Force Generation

As a result of the ATP hydrolysis, filaments elongate approximately 10 times faster at their barbed ends than their pointed ends. At steady-state, the polymerization rate at the barbed end matches the depolymerization rate at the pointed end, and microfilaments are said to be treadmilling. Treadmilling results in elongation in the barbed end and shortening in the pointed-end, so that the filament in total moves. Since both processes are energetically favourable, this means force is generated, and the energy ultimately comes from the NTP hydrolization.

Actin in Cells

Intracellular actin cytoskeletal assembly and disassembly are tightly regulated by cell signalling mechanisms. Many signal transduction systems use the actin cytoskeleton as a scaffold holding them at or near the inner face of the peripheral membrane.

This subcellular location allows immediate and exquisite responsiveness to transmembrane receptor action and signal-processing enzyme cascades. Because actin monomers must be recycled to sustain high rates of actin-based motility during chemotaxis, cell signalling is believed to activate cofilin, an actin-filament depolymerizing protein which binds to ADP-rich actin subunits nearest the filament's pointed-end and promotes filament fragmentation, with concomitant depolymerization to liberate actin monomers. The protein profilin enhances the ability of monomers to assemble by stimulating the exchange of actin-bound ADP for solution-phase ATP to yield Actin-ATP and ADP. In most animal cells, monomeric actin is bound to profilin and thymosin-beta4, both of which preferentially bind with one-to-one stoichiometry to ATP-containing monomers.

Although thymosin-beta4 is strictly a monomer-sequestering protein, the behaviour of profilin is far more complex. Profilin is transferred to the leading edge by virtue of its PIP2 binding site, and profilin also employs its poly-L-proline binding site to dock onto end-tracking proteins. Once bound, Profilin-Actin-ATP is loaded into the monomer-insertion site of actoclampin motors. Another important component in filament formation is the Arp2/3 complex, which binds to the side of an already existing filament (or "mother filament"), where it nucleates the formation of a new actin filament and creates a fan-like branched filament network.

In non-muscle cells, actin filaments are formed at/near membrane surfaces. Their formation and turnover are regulated by many proteins, including

- Filament end-tracking protein (e.g., formins, VASP, N-WASP)
- Filament-nucleator known as the Actin-Related Protein-2/3 (or Arp2/3) complex
- Filament cross-linkers (e.g., a-actinin and fascin)
- Actin monomer-binding proteins profilin and thymosin-a4
- Filament barbed-end cappers such as Capping Protein and CapG, *etc.*
- Filament-severing proteins like gelsolin
- pointed-End depolymerizing proteins such as ADF/cofilin.

The actin filament network in non-muscle cells is highly dynamic. As first proposed by Dickinson & Purich (*Biophysical Journal* 92: 622-631), the actin filament network is arranged with the barbed-end of each filament attached to the cell's peripheral membrane by means of clamped-filament elongation motors ("actoclampins") formed from a filament barbed-end and a clamping protein (formins, VASP, Mena, WASP, and N-WASP). The primary substrate for these elongation motors is Profilin-Actin-ATP complex which is directly transferred to elongating filament ends (Dickinson, Southwick & Purich, 2002). The pointed-end of each filament is oriented toward the cell's interior. In the case of lamellipodial growth, the Arp2/3 complex generates a branched network, and in filopodia a parallel array of filaments is formed.

Actin Acts as a Track for Myosin Motor Motility

Myosin motors are intracellular ATP-dependent enzymes that bind to and move along actin filaments. Various classes of myosin motors have very different behaviors, including exerting tension in the cell and transporting cargo vesicles.

A Proposed Model - Actoclampins Track Filament Ends

One proposed model suggests the existence of actin filament barbed-end-tracking molecular motors termed actoclampin. The proposed actoclampins generate the propulsive forces needed for actin-based motility of lamellipodia, filopodia, invadipodia, dendritic spines, intracellular vesicles, and motile processes in endocytosis, exocytosis, podosome formation, and phagocytosis. Actoclampin motors also propel such intracellular pathogens as *Listeria monocytogenes*, *Shigella flexneri*, *Vaccinia* and *Rickettsia*. When assembled under suitable conditions, these end-tracking molecular motors can also propel biomimetic particles.

The term actoclampin is derived from *acto-* to indicate the involvement of an actin filament, as in actomyosin, and *clamp* to indicate a clasping device used for strengthening flexible/moving objects and for securely fastening two or more components, followed by the suffix *-in* to indicate its protein origin. An actin filament end-tracking protein may thus be termed a clampin.

Dickinson and Purich (2002) recognized that prompt ATP hydrolysis could explain the forces achieved during actin-based motility. They proposed a simple mechanoenzymatic sequence known as the Lock, Load & Fire Model, in which an end-tracking protein remains tightly bound ("locked" or clamped) onto the end of one sub-filament of the double-stranded actin filament. After binding to Glycyl-Prolyl-Prolyl-Prolyl-Prolyl-Prolyl-registers on tracker proteins, Profilin-ATP-actin is delivered ("loaded") to the unclamped end of the other sub-filament, whereupon ATP within the already clamped terminal subunit of the other subfragment is hydrolyzed ("fired"), providing the energy needed to release that arm of the end-tracker, which then can bind another Profilin-ATP-actin to begin a new monomer-addition round.

The following steps describe one force-generating cycle of an actoclampin molecular motor:

1. The polymerization cofactor profilin and the ATP·actin combine to form a profilin-ATP-actin complex that then binds to the end-tracking unit
2. The cofactor and monomer are transferred to the barbed-end of an actin already clamped filament
3. The tracking unit and cofactor dissociate from the adjacent protofilament, in a step that can be facilitated by ATP hydrolysis energy to modulate the affinity of the cofactor and/or the tracking unit for the filament; and this mechanoenzymatic cycle is then repeated, starting this time on the other sub-filament growth site.

When operating with the benefit of ATP hydrolysis, AC motors generate per-filament forces of 8–9 pN, which is far greater than the per-filament limit of 1–2 pN for motors operating without ATP hydrolysis (Dickinson and Purich, 2002, 2006; Dickinson, Caro and Purich, 2004). The term actoclampin is generic and applies to all actin filament end-tracking molecular motors, irrespective of whether they are driven actively by an ATP-activated mechanism or passively.

Some actoclampins (e.g., those involving Ena/VASP proteins, WASP, and N-WASP) apparently require Arp2/3-mediated filament initiation to form the actin polymerization nucleus that is then "loaded" onto the end-tracker before processive motility can commence. To generate a new filament, Arp2/3 requires a "mother" filament, monomeric ATP-actin, and an activating domain from Listeria ActA or the VCA region of N-WASP. Ther Arp2/3 complex binds to the side of the mother filament, forming a Y-shaped branch having a 70 degree angle with respect to the longitudinal axis of the mother filament. Then upon activation by ActA or VCA, the Arp complex is believed to undergo a major conformational change, bringing its two actin-related protein subunits near enough to each other to generate a new filament gat. Whether ATP hydrolysis may be required for nucleation and/or Y-branch release is a matter under active investigation.

Actin

Actin is a globular, roughly 42-kDa protein found in all eukaryotic cells (the only known exception being nematode sperm) where it may be present at concentrations of over 100 μM. It is also one of the most highly-conserved proteins, differing by no more than 20% in species as diverse as algae and humans. Actin is the monomeric subunit of two types of filaments in cells: microfilaments, one of the three major components of the cytoskeleton, and thin filaments, part of the contractile apparatus in muscle cells. Thus, actin participates in many important cellular processes including muscle contraction, cell motility, cell division and cytokinesis, vesicle and organelle movement, cell signalling, and the establishment and maintenance of cell junctions and cell shape. Many of these processes are mediated by extensive and intimate interactions of actin with cellular membranes. In vertebrates, three main groups of actin isoforms, alpha, beta, and gamma have been identified. The alpha actins are found in muscle tissues and are a major constituent of the contractile apparatus. The beta and gamma actins co-exist in most cell types as components of the cytoskeleton, and as mediators of internal cell motility.

Formation of Thin Filament

Principal interactions of structural proteins are at cadherin-based adherens junction. Actin filaments are linked to α-actinin and to the membrane through vinculin. The head domain of vinculin associates to E-cadherin via á-, â-, and β-catenins. The tail domain of vinculin binds to membrane lipids and to actin filaments. The protein actin is one of the most highly conserved throughout evolution because it interacts with a large number of other proteins, with 80.2% sequence conservation at the gene level between *Homo sapiens* and *Saccharomyces cerevisiae* (a species of yeast), and 95% conservation of the primary structure of the protein product.

Although most yeasts have only a single actin gene, higher eukaryotes, in general, express several isoforms of actin encoded by a family of related genes. Mammals have at least six actin isoforms coded by separate genes, which are divided into three classes (alpha, beta and gamma) according to their isoelectric point. In general, alpha actins are found in muscle, whereas beta and gamma isoforms are prominent in non-muscle cells (α- and β1-cytoplasmic).

Although the amino acid sequences and *in vitro* properties of the isoforms are highly similar, these isoforms cannot completely substitute for one another *in vivo*.

The typical actin gene has an approximately 100-nucleotide 5' UTR, a 1200-nucleotide translated region, and a 200-nucleotide 3' UTR. The majority of actin genes are interrupted by introns, with up to 6 introns in any of 19 well-characterised locations. The high conservation of the family makes actin the favoured model for studies comparing the introns-early and introns-late models of intron evolution.

All non-spherical prokaryotes appear to possess genes such as MreB, which encode homologues of actin; these genes are required for the cell's shape to be maintained. The plasmid-derived gene ParM encodes an actin-like protein whose polymerised form is dynamically unstable, and appears to partition the plasmid DNA into the daughter cells during cell division by a mechanism analogous to that employed by microtubules in eukaryotic mitosis. Actin is found in both smooth and rough endoplasmic reticulums.

Functions

Actin has four main functions in cells:

- To form microfilaments, the most dynamic of the three subclasses of the cytoskeleton, which give mechanical support to cells, and hardwire the cytoplasm with the surroundings to support signal transduction.
- To allow cell motility, including phagocytosis of bacteria by macrophages.
- In muscle cells, to be the scaffold on which myosin proteins generate force to support muscle contraction.
- In non-muscle cells, to be a track for cargo transport myosins (non-conventional myosins) such as myosin V and VI. Non-conventional myosins use ATP hydrolysis to transport cargo, such as vesicles and organelles, in a directed fashion much faster than diffusion. Myosin V walks towards the barbed end of actin filaments, while myosin VI walks toward the pointed end. Most actin filaments are arranged with the barbed end toward the cellular membrane and the pointed end toward the cellular interior. This arrangement allows myosin V to be an effective motor for export of cargos, and myosin VI to be an effective motor for import.

Directionality

The polarity of an actin filament can be determined by decorating the microfilament with myosin "S1" fragments, creating barbed (+) and pointed (-) ends on the filament. An S1 fragment is composed of the head and neck domains of myosin II. Under physiologic conditions, G-actin (the monomer form) is transformed to F-actin (the polymer form) by ATP, where the role of ATP is essential.

Nucleation and Polymerization

Actin polymerization and depolymerization is necessary in chemotaxis and cytokinesis. Nucleating factors are necessary to stimulate actin polymerization. One such nucleating factor is the ARP complex which acts as a barbed end of actin in its shape to stimulate the nucleation of G-actin (or monomeric actin). The Arp2/3 complex can also bind to actin filaments at 70 degrees to form new actin branches off of existing actin filaments. Also, actin filaments themselves bind ATP, and hydrolysis of this ATP stimulates destabilization of the polymer.

The growth of actin filaments can be regulated by thymosin and profilin. Thymosin binds to G-actin to prevent it from polymerizing while profilin binds to G-actin to promote monomeric addition to the barbed, plus end.

Microfilaments

Individual subunits of microfilaments are known as globular actin (G-actin). G-actin subunits assemble into long filamentous polymers called F-actin. Two parallel F-actin strands must rotate 166 degrees to layer correctly on top of each other. This creates the double helix structure of the microfilaments of the cytoskeleton. Microfilaments measure approximately 7 nm in diameter with a loop of the helix repeating every 37 nm.

Actomyosin Filaments

In muscle, actin is the major component of *thin filaments*, which, together with the motor protein myosin (which forms *thick filaments*), are arranged into actomyosin myofibrils. These fibrils comprise the mechanism of muscle contraction. Using the hydrolysis of ATP for energy, myosin heads undergo a cycle during which they attach to thin filaments, exerting a tension, and then depending on the load, perform a power stroke that causes the thin filaments to slide past, shortening the muscle.

In contractile bundles, the actin-bundling protein alpha-actinin separates each thin filament by ~35 nm. This increase in distance allows thick filaments to fit in between and interact, enabling deformation or contraction. In deformation, one end of myosin is bound to the plasma membrane while the other end "walks" toward the plus end of the actin filament. This pulls the membrane into a different shape relative to the cell cortex.

For contraction, the myosin molecule is usually bound to two separate filaments and both ends simultaneously "walk" towards their filament's plus end, sliding the actin filaments closer to each other. This results in the shortening, or contraction, of the actin bundle (but not the filament). This mechanism is responsible for muscle contraction and cytokinesis, the division of one cell into two.

Nuclear Actin

Actin is essential for transcription from RNA polymerases I, II and III. In Pol I transcription, actin and myosin (MYO1C, which binds DNA) act as a molecular motor. For Pol II transcription, â-actin is needed for the formation of the pre-initiation complex. Pol III contains α-actin as a subunit. Actin can also be a component of chromatin remodeling complexes as well as pre-mRNP particles (that is, precursor messenger RNA bundled in proteins), and is involved in nuclear export of RNAs and proteins.

History

Actin was first observed experimentally in 1887 by W.D. Halliburton, who extracted a protein from muscle that 'coagulated' preparations of myosin, which he dubbed "myosin-ferment." However, Halliburton was unable to further characterise his findings, and the discovery of actin is credited instead to Brúnó F. Straub, a young biochemist working in Albert Szent-Györgyi's laboratory at the Institute of Medical Chemistry at the University of Szeged, Hungary. In 1942, Straub developed a novel technique for extracting muscle protein that allowed him to isolate substantial amounts of relatively pure actin. Straub's method is essentially the same as that used in laboratories today.

Szent-Gyorgyi had previously described the more viscous form of myosin produced by slow muscle extractions as 'activated' myosin, and, since Straub's protein produced the activating effect, it was dubbed *actin*. The hostilities of World War II meant that Szent-Gyorgyi and Straub were unable to publish the work in Western scientific journals; it became well-known in the West only in 1945, when it was published as a supplement to the Acta Physiologica Scandinavica.

Straub continued to work on actin, and in 1950 reported that actin contains bound ATP and that, during polymerisation of the protein into microfilaments, the nucleotide is hydrolysed to ADP and inorganic phosphate (which remain bound in the microfilament). Straub suggested that the transformation of ATP-bound actin to ADP-bound actin played a role in muscular contraction. In fact, this is true only in smooth muscle, and was not supported through experimentation until 2001. The crystal structure of G-actin was solved in 1990 by Kabsch and colleagues. In the same year, a model for F-actin was proposed by Holmes and colleagues. The model was derived by fitting a helix of G-actin structures according to low-resolution fibre diffraction data from the filament. Several models of the filament have been proposed since. However, there is still no high-resolution X-ray structure of F-actin. The Listeria bacteria use the cellular machinery to move around inside the host cell, by inducing directed polymerisation of actin by the ActA transmembrane protein, thus pushing the bacterial cell around.

Intermediate Filaments

Intermediate filaments (IFs) are a family of related proteins that share common structural and sequence features. Intermediate filaments have an average diameter of 10 nanometers, which is between that of actin (microfilaments) and microtubules, although they were initially designated 'intermediate' because their average diameter was between those of narrower microfilaments (actin) and wider myosin filaments. Most types of intermediate filaments are cytoplasmic, but one type, the lamins, are nuclear.

Structure

The domain structure of IF molecules is conserved. Each protein has a non-alpha-helical (globular) domain at the N and C-termini which surrounds the alpha-helical rod domain. The basic building block for IFs is a parallel and in register dimer. The dimer is formed through the interaction of the rod domain to form a coiled coil. Cytoplasmic IF assemble into non-polar unit-length filaments (ULF) which then assemble into longer structures. Part of the assembly process includes a compaction step, in which ULF tighten and assume a smaller diameter. The reasons for this compaction are not well understood, and IF are routinely observed to have diameters ranging between 6 and 12nm. The anti-parallel orientation of tetramers means that, unlike microtubules and microfilaments which have a plus end and a minus end, IFs lack polarity. Also, as opposed to actin or tubulin, intermediate filaments do not contain a binding site for a nucleoside triphosphate. Cytoplasmic IF do not undergo treadmilling like microtubules and actin fibres, but they are dynamic.

Biomechanical Properties

IFs are rather deformable proteins that can be stretched several times their initial length. The key to facilitate this large deformation is due to their hierarchical structure, which facilitates a cascaded activation of deformation mechanisms at different levels of strain.

Types

There are about 70 different genes coding for various intermediate filament proteins. However, different kinds of IFs share basic characteristics: they are all polymers that generally measure between 9-11 nm in diameter when fully assembled. IF are subcategorized into six types based on similarities in amino acid sequence and protein structure.

Types I and II - Acidic and Basic Keratins

These proteins are the most diverse among IFs and constitute type I (acidic) and type II (basic) IF proteins. The many isoforms are divided in two groups:

- epithelial keratins (about 20) in epithelial cells (image to right)
- trichocytic keratins (about 13) (hair keratins) which make up hair, nails, horns and reptilian scales.

Regardless of the group, keratins are either acidic or basic. Acidic and basic keratins bind each other to form acidic-basic heterodimers and these heterodimers then associate to make a keratin filament.

Type III

There are four proteins classed as type III IF proteins which may form homo- or heteropolymeric proteins.

- Desmin IFs are structural components of the sarcomeres in muscle cells.
- GFAP (glial fibrillary acidic protein) is found in astrocytes and other glia.
- Peripherin found in peripheral neurons.
- Vimentin, the most widely distributed of all IF proteins, can be found in fibroblasts, leukocytes, and blood vessel endothelial cells. They support the cellular membranes and keep some organelles in a fixed place within the cytoplasm.

Type IV

- α-Internexin
- Neurofilaments - the type IV family of intermediate filaments that is found in high concentrations along the axons of vertebrate neurons.
- Synemin
- Syncoilin.

Type V - Nuclear Lamins

Lamins : Lamins are fibrous proteins having structural function in the cell nucleus. In metazoan cells there are A and B type lamins which differ in their length and pI. Human cells have three differentially regulated genes. B-type lamins are present in every cell. B type lamins, B1 and B2, are expressed from the LMNB1 and LMNB2 genes on 5q23 and 19q13, respectively. A-type lamins are only expressed following gastrulation. Lamin A and C are the most common A-type lamins and are splice variants of the LMNA gene found at 1q21.

These proteins localize to two regions of the nuclear compartment, the nuclear lamina — a proteinaceous structure layer subjacent to the inner surface of the nuclear envelope and throughout the nucleoplasm in the nucleoplasmic "veil". Comparison of the lamins to vertebrate cytoskeletal IFs shows that lamins have an extra 42 residues (six heptads) within coil 1b. The c-terminal tail domain contains a nuclear localization signal (NLS), an Ig-fold like domain, and in most cases a carboxy-terminal CaaX box that is isoprenylated and carboxymethylated (lamin C does not have a CAAX box). Lamin A is further processed to remove the last 15 amino acids and its farnesylated cysteine. During mitosis, lamins are phosphorylated by MPF which drives the disassembly of the lamina and the nuclear envelope.

Cell Adhesion

At the plasma membrane, some keratins interact with desmosomes (cell-cell adhesion) and hemidesmosomes (cell-matrix adhesion) via adapter proteins.

Associated Proteins

Filaggrin binds to keratin fibres in epidermal cells. Plectin links vimentin to other vimentin fibres, as well as to microfilaments, microtubules, and myosin II. Keratin filaments in epithelial cells link to desmosomes (desmosomes connect the cytoskeleton together) through plakoglobin, desmoplakin, desmogleins and desmocollins; desmin filaments are connected in a similar way in heart muscle cells.

Diseases Arising from Mutations in IF Genes

- Epidermolysis bullosa simplex; K5 or K14 mutation
- Laminopathies are a family of diseases caused by mutations in nuclear lamins and include Hutchinson Gilford Progeria Syndrome and various lipodystrophies and cardiomyopathies among others.
- Human Intermediate Filament Database(HIFD), a comprehensive database of human intermediate filament proteins, their associated variations and diseases.
- envelope.

Microtubules

Microtubules are one of the components of the cytoskeleton. They have a diameter of 25 nm and length varying from 200 nanometers to 25 micrometers. Microtubules serve as structural components within cells and are involved in many cellular processes including mitosis, cytokinesis, and vesicular transport.

Structure

Microtubules are polymers of α- and β-tubulin dimers. The tubulin dimers polymerize end to end in protofilaments. The protofilaments then bundle into hollow cylindrical filaments. Typically the protofilaments arrange themselves in an imperfect helix with one turn of the helix containing 13 tubulin dimers each from a different protofilament. The image above illustrates a small section of microtubule, a few αβ dimers in length. Another important feature of microtubule structure is polarity. Tubulin polymerizes end to end with the α subunit of one tubulin dimer contacting the β subunit of the next. Therefore, in a protofilament, one end will have the α subunit exposed while the other end will have the α subunit exposed. These ends are designated the (") and (+) ends, respectively. The protofilaments bundle parallel to one another, so in a microtubule, there is one end, the (+) end, with only α subunits exposed while the other end, the (") end, only has β subunits exposed. The (-) end is capped so elongation of the microtubule occurs from the (+) direction.

Organization within Cells

Microtubules are nucleated and organized by the microtubule organizing centres (MTOCs), such as centrioles and basal bodies. Microtubules are part of a structural network (the cytoskeleton) within the cell's cytoplasm, but, in addition to structural support, microtubules take part in many other processes, as well. They are capable of growing and shrinking in order to generate force, and there are also motor proteins that allow organelles and other cellular factors to move along the microtubule. A notable structure involving microtubules is the mitotic spindle used by eukaryotic cells to segregate their chromosomes correctly during cell division. Microtubules are also part of the cilia and flagella of eukaryotic cells (prokaryote flagella are entirely different). The process of mitosis is facilitated by a subgroup of microtubules known as astral microtubules, defined as a microtubule originating from the centrosome that does not connect to a kinetochore. Astral microtubules develop in the actin skeleton and interact with the cell cortex to aid in spindle orientation.

They are organized into radial arrays around the centrosomes. The turnover rate of this population of microtubules is higher than any other population. Astral microtubules function in concert with specialized dynein motors, which are oriented with the light chain portion attached to the cell membrane and the dynamic portion attached to the microtubule. This allows for dynein contraction to pull the centrosome towards the cell membrane, thus assisting in cytokinesis. Astral microtubules are not required for the progression of mitosis, but they are required to ensure the fidelity of the process; they are required for the correct positioning and orientation of the mitotic spindle apparatus. They are also involved in determination of cell division site based on the geometry and polarity of the cells. The maintenance of astral microtubules is dependent on the integrity of centrosome. It is also dependent on several microtubule-associated proteins such as EB1 and Adenomatous Polyposis Coli (APC).

Nucleation and Growth

Polymerization of microtubules is nucleated in a microtubule organizing centre. Contained within the MTOC is another type of tubulin, β-tubulin, which is distinct from the α and β subunits which compose the microtubules themselves. The α-tubulin combines with several other associated proteins to form a circular structure known as the "α-tubulin ring complex." This complex acts as a scaffold for α/β tubulin dimers to begin polymerization; it acts as a cap of the (") end while microtubule growth continues away from the MTOC in the (+) direction.

Dynamic Instability

During polymerization, both the α- and β-subunits of the tubulin dimer are bound to a molecule of GTP. While the GTP bound to α-tubulin is stable, the GTP bound to α-tubulin may be hydrolyzed to GDP shortly after assembly. The kinetics of GDP-tubulin are different from those of GTP-tubulin; GDP-tubulin is prone to depolymerization. A GDP-bound tubulin subunit at the tip of a microtubule will fall off, though a GDP-bound tubulin in the middle of a microtubule cannot spontaneously pop out. Since tubulin adds onto the end of the microtubule only in the GTP-bound state, there is generally a cap of GTP-bound tubulin at the tip of the microtubule, protecting it from disassembly. When hydrolysis catches up to the tip of the microtubule, it begins a rapid depolymerization and shrinkage. This switch from growth to shrinking is called a catastrophe. GTP-bound tubulin can begin adding to the tip of the microtubule again, providing a new cap and protecting the microtubule from shrinking. This is referred to as rescue.

Chemical Effects on Microtubule Dynamics

Microtubule dynamics can also be altered by drugs.

- For example, the taxane drug class (e.g. paclitaxel or docetaxel), used in the treatment of cancer, blocks dynamic instability by stabilizing GDP-bound tubulin in the microtubule. Thus, even when hydrolysis of GTP reaches the tip of the microtubule, there is no depolymerization and the microtubule does not shrink back.
- Nocodazole and Colchicine have the opposite effect, blocking the polymerization of tubulin into microtubules.

Motor Proteins

In addition to movement generated by the dynamic instability of the microtubule itself, the fibres are substrates along which motor proteins can move. The major microtubule motor proteins are kinesin, which generally moves towards the (+) end of the microtubule, and dynein, which generally moves towards the (") end.

Postulated Role in Consciousness

In their controversial Orch-OR theory of consciousness, Roger Penrose and Stuart Hameroff postulate that microtubules in neurons conduct quantum-level manipulations of matter which produces consciousness, based partially on widely recorded observations of Gamma Synchrony that indicate that information may propagate through the brain much faster than a chemically mediated neural network would physically permit. Max Tegmark disputes the relevance of these observations, and the matter remains open to debate.

The Prokaryotic Cytoskeleton

The prokaryotic cytoskeleton is the collective name for all structural filaments in prokaryotes. It was once thought that prokaryotic cells did not possess cytoskeletons, but recent advances in visualization technology and structure determination have shown that filaments indeed exist in these cells. In fact, analogues for all major cytoskeletal proteins in eukaryotes have been found in prokaryotes. Cytoskeletal elements play essential roles in cell division, protection, shape determination, and polarity determination in various prokaryotes.

The cytoskeleton was previously thought to be a feature only of eukaryotic cells, but homologues to all the major proteins of the eukaryotic cytoskeleton have recently been found in prokaryotes. Although the evolutionary relationships are so distant that they are not obvious from protein sequence comparisons alone, the similarity of their three-dimensional structures and similar functions in maintaining cell shape and polarity provides strong evidence that the eukaryotic and prokaryotic cytoskeletons are truly homologous. However, some structures in the bacterial cytoskeleton may have yet to be identified.

FtsZ, the first identified prokaryotic cytoskeletal element, forms a filamentous ring structure located in the middle of the cell called the Z-ring that constricts during cell division, similar to the actin-myosin contractile ring in eukaryotes. The Z-ring is a highly dynamic structure that consists of numerous bundles of protofilaments that extend and shrink, although the mechanism behind Z-ring contraction and the number of protofilaments involved are unclear. FtsZ acts as an organizer protein and is required for cell division. It is the first component of the septum during cytokinesis, and it recruits all other known cell division proteins to the division site.

Despite this functional similarity to actin, FtsZ is homologous to eukaryal tubulin. Although comparison of the primary structures of FtsZ and tubulin reveal a weak relationship, their 3-dimensional structures are remarkably similar. Furthermore, like tubulin, monomeric FtsZ is bound to GTP and polymerizes with other FtsZ monomers with the hydrolysis of GTP in a mechanism similar to tubulin dimerization. Since FtsZ is essential for cell division in bacteria, this protein is a target for the design of new antibiotics.

MreB

MreB is a bacterial protein believed to be analogous to eukaryal actin. MreB and actin have a weak primary structure match, but are very similar in terms of 3-D structure and filament polymerization.

Almost all non-spherical bacteria rely on MreB to determine their shape. MreB assembles into a helical network of filamentous structures just under the cytoplasmic membrane, covering the whole length of the cell. MreB determines cell shape by mediating the position and activity of enzymes that synthesize peptidoglycan and by acting as a rigid filament under the cell membrane that exerts outward pressure to sculpt and bolster the cell. MreB condenses from its normal helical network and forms a tight ring at the septum in *Caulobacter crescentus* right before cell division, a mechanism that is believed to help locate its off-centre septum. MreB is also important for polarity determination in polar bacteria, as it is responsible for the correct positioning of at least four different polar proteins in *C. crescentus*.

Crescentin

Crescentin (encoded by *creS* gene) is an analogue of eukaryotic intermediate filaments (IFs). Unlike the other analogous relationships discussed here, crescentin has a rather large primary homology with IF proteins in addition to three-dimensional similarity - the sequence of *creS* has a 25% identity match and 40% similarity to cytokeratin 19 and a 24% identity match and 40% similarity to nuclear lamin A. Furthermore, crescentin filaments are roughly 10 nm in diameter and thus fall within diameter range for eukaryal IFs (8-15 nm). Crescentin forms a continuous filament from pole to pole alongside the inner, concave side of the crescent-shaped bacterium *Caulobacter crescentus*. Both MreB and crescentin are necessary for *C. crescentus* to exist in its characteristic shape; it is believed that MreB molds the cell into a rod shape and crescentin bends this shape into a crescent.

ParM and SopA

ParM is a cytoskeletal element that possesses a similar structure to actin, although it behaves functionally like tubulin. Further, it polymerizes bidirectionally and it exhibits dynamic instability, which are both behaviors characteristic of tubulin polymerization. It forms a system with ParR and *parC* that is responsible for R1 plasmid separation. ParM affixes to ParR, a DNA-binding protein that specifically binds to 10 direct repeats in the *parC* region on the R1 plasmid. This binding occurs on both ends of the ParM filament. This filament is then extended, separating the plasmids. The system is analogous to eukaryotic chromosome segregation as ParM acts like eukaryotic tubulin in the mitotic spindle, ParR acts like the kinetochore complex, and *parC* acts like the centromere of the chromosome. F plasmid segregation occurs in a similar system where SopA acts as the cytoskeletal filament and SopB binds to the *sopC* sequence in the F plasmid, like the kinetochore and centromere respectively.

MinCDE System

The MinCDE system is a filament system that properly positions the septum in the middle of the cell in *Escherichia coli*. According to Shih et al., MinC inhibits the formation of the septum by prohibiting the polymerization of the Z-ring. MinC, MinD, and MinE form a helix structure that winds around the cell and is bound to the membrane by MinD. The MinCDE helix occupies a pole and terminates in a filamentous structure called the E-ring made of MinE at the middle-most edge of the polar zone.

From this configuration, the E-ring will contract and move towards that pole, disassembling the MinCDE helix as it moves along. Concomitantly, the disassembled fragments will reassemble at the opposite polar end, reforming the MinCDE coil on the opposite pole while the current MinCDE helix is broken down. This process then repeats, with the MinCDE helix oscillating from pole to pole. This oscillation occurs repeatedly during the cell cycle, thereby keeping MinC (and its septum inhibiting effect) at a lower time-averaged concentration at the middle of the cell than at the ends of the cell.

The dynamic behaviour of the Min proteins has been reconstituted in vitro using an artificial lipid bilayer as mimic for the cell membrane. MinE and MinD self-organized into parallel and spiral protein waves by a reaction-diffusion like mechanism.

Microtrabeculae

A fourth eukaryotic cytoskeletal element, *microtrabeculae*, was proposed by Keith Porter based on images obtained from high-voltage electron microscopy of whole cells in the 1970s. The images showed short, filamentous structures of unknown molecular composition associated with known cytoplasmic structures.

Porter proposed that this microtrabecular structure represented a novel filamentous network distinct from microtubules, filamentous actin, or intermediate filaments. It is now generally accepted that microtrabeculae are nothing more than an artifact of certain types of fixation treatment, although we have yet to fully understand the complexity of the cell's cytoskeleton.

Cilia and Flagella

A cilium (plural *cilia*) is an organelle found in eukaryotic cells. Cilia are slender protuberances that project from the much larger cell body. There are two types of cilia: *motile* cilia and *non-motile*, or primary cilia, which typically serve as sensory organelles. In eukaryotes, cilia and flagella together make up a group of organelles known as undulipodia. Eukaryotic cilia are structurally identical to Eukaryotic flagella, although distinctions are sometimes made according to function and/or length.

Types and Distribution

In humans, *primary* cilia are found on nearly every cell in the body. Cilia are rare in most plants, occurring most notably in cycads. Ciliates are microscopic organisms that possess *motile* cilia exclusively and use them for either locomotion or to simply move liquid over their surface. Larger eukaryotes, such as mammals, have *motile* cilia as well. Motile cilia are usually present on a cell's surface in large numbers and beat in coordinated waves. In humans, for example, motile cilia are found in the lining of the trachea (windpipe), where they sweep mucus and dirt out of the lungs. In female mammals, the beating of cilia in the Fallopian tubes moves the ovum from the ovary to the uterus.

In comparison to *motile* cilia, non-motile (or *primary*) cilia usually occur one per cell; nearly all mammalian cells have a single non-motile *primary cilium*. In addition, examples of specialized primary cilia can be found in human sensory organs such as the eye and the nose. The outer segment of the rod photoreceptor cell in the human eye is connected to its cell body with a specialized non-motile cilium. The dendritic knob of the olfactory neuron, where the odorant receptors are located, also contains non-motile cilia (about 10 cilia per dendritic knob).

Although the primary cilium was discovered in 1898, it was largely ignored for a century. Only recently has great progress been made in understanding the function of the primary cilium. Until the 1990s, the prevailing view of the primary cilium was that it was merely a vestigial organelle, without important function. Recent findings regarding its physiological roles in chemical sensation, signal transduction, and control of cell growth, have led scientists to acknowledge its importance in cell function, with the discovery of its role in diseases not previously recognized to involve the dysgenesis and dysfunction of cilia, such as polycystic kidney disease congenital heart disease, and an emerging group of genetic ciliopathies. The primary cilium is now known to play an important role in the function of many human organs. The current scientific understanding of primary cilia views them as "sensory cellular antennae that coordinate a large number of cellular signalling pathways, sometimes coupling the signalling to ciliary motility or alternatively to cell division and differentiation."

Structure, Assembly and Maintenance, and Function

"Inside cilia and flagella is a microtubule-based cytoskeleton called the axoneme. The axoneme of primary cilia typically has a ring of nine outer microtubule doublets (called a 9+0 axoneme), and the axoneme of a motile cilium has two central microtubule singlets in addition to the nine outer doublets (called a 9+2 axoneme). The axonemal cytoskeleton acts as a scaffolding for various protein complexes and provides binding sites for molecular motor proteins such as kinesin II, that help carry proteins up and down the microtubules."

The building blocks of the cilia such as tubulins and other partially assembled axonemal proteins are added to the ciliary tips which point away from the cell body. In most species bi-directional motility called intraflagellar transport or IFT plays an essential role to move these building materials from the cell body to the assembly site. IFT also carries the disassembled material to be recycled from the ciliary tip back to the cell body. By regulating the equilibrium between these two IFT processes, the length of cilia can be maintained dynamically. The disassembly of the cilia requires the action of the protein kinase Aurora A.

Exceptions where IFT is not present include *Plasmodium falciparum* which is one of the species of *Plasmodium* that cause malaria in humans. In this parasite, cilia assemble in the cytoplasm.

At the base of the cilium where the cilia attaches to the cell body is the microtubule organizing centre, the basal body. Some basal body proteins as CEP164, ODF2 and CEP170, regulate the formation and the stability of the cilium. A transition zone between the basal body and the axoneme "serves as a docking station for intraflagellar transport and motor proteins."

> *In effect, the [cilium] is a nanomachine composed of perhaps over 600 proteins in molecular complexes, many of which also function independently as nanomachines.*

Sensing the Extracellular Environment

Some epithelial cells are ciliated, and they commonly exist as a sheet of polarised cells forming a tube or tubule with cilia projecting into the lumen." Primary cilia on epithelial cells provide chemosensation, thermosensation and mechanosensation of the extracellular environment by playing "a sensory role mediating specific signalling cues, including soluble factors in the external cell environment, a secretory role in which a soluble protein is released to have an effect downstream of the fluid flow, and mediation of fluid flow if the cilia are motile.

Cilium-related Disease

Ciliary defects can lead to a number of human diseases. Genetic mutations compromising the proper functioning of cilia, ciliopathies, can cause chronic disorders such as primary ciliary dyskinesia (PCD), nephronophthisis or Senior-Loken syndrome. In addition, a defect of the primary cilium in the renal tube cells can lead to polycystic kidney disease (PKD). In another genetic disorder called Bardet-Biedl syndrome (BBS), the mutant gene products are the components in the basal body and cilia. Lack of functional cilia in female

Fallopian tubes can cause ectopic pregnancy. A fertilized ovum may not reach the uterus if the cilia are unable to move it there. In such a case, the ovum will implant in the Fallopian tubes, causing a tubal pregnancy, the most common form of ectopic pregnancy.

Since the flagellum of human sperm is actually a modified cilium, ciliary dysfunction can also be responsible for male infertility. Of interest, there is an association of primary ciliary dyskinesia with left-right anatomic abnormalities such as situs inversus (a combination of findings known as Kartagener's syndrome) and other heterotaxic defects. These left-right anatomic abnormalities can also result in congenital heart disease. In fact, it has been shown that proper cilial function is responsible for the normal left-right asymmetry in mammals.

Ciliopathy as an Origin for Many Multi-symptom Genetic Diseases

Recent findings in genetic research have suggested that a large number of genetic disorders, both genetic syndromes and genetic diseases, that were not previously related in the medical literature, may be, in fact, highly related in the root cause of the widely-varying set of medical symptoms that are clinically visible in the disorder. These have been grouped as an emerging class of diseases called ciliopathies. The underlying cause may be a dysfunctional molecular mechanism in the primary cilia structures, organelles which are present in many diverse cellular types throughout the human body. Cilia defects adversely affect "numerous critical developmental signalling pathways" essential to cellular development and thus offer a plausible hypothesis for the often multi-symptom nature of a large set of syndromes and diseases. Known ciliopathies include primary ciliary dyskinesia, Bardet-Biedl syndrome, polycystic kidney and liver disease, nephronophthisis, Alstrom syndrome, Meckel-Gruber syndrome and some forms of retinal degeneration.

Mucociliary Clearance

Mucociliary clearance, also referred to as mucociliary apparatus or mucociliar clearance (MCC), derived from *mucus*, *cilia* (cilia of the tracheal surface epithelium in the respiratory tract) and *clearance* describes the self-clearing mechanism of the bronchi. The main bronchi down to the alveoli are lined with a respiratory epithelium. On that, cilium is present, bearing hair-shaped structures on its surface (cilia). The cilia are surrounded by a thin fluid film of mucus. On top of that is a second viscous film of mucus, in which foreign particles and microorganisms get stuck. Within the thin fluid film of mucus the cilia act out movements coordinated in direction towards the pharynx. Thereby the viscous film of mucus including its freight is transported off in direction towards the mouth, where it is either swallowed or expelled via coughing.

Important for good mucociliary clearance are the number of cilia, their structure, activity, and coordinated movement. Optimum functionality of mucociliary clearance presuppose a temperature of 37 °C and an absolute humidity of 44 mg/dm^3 corresponding to a relative humidity of 100 %. Under the condition of insufficient temperature and humidity, after a short time the ciliary cells suspend their transport function. Under such circumstances, bacterial germinal colonization is facilitated. Pulmonary infections and damaging of the pulmonic tissues may be the consequence. High humidity enhances the functionality of the mucociliary clearance. Two methods of supporting mucociliary clearance are the active and the passive respiratory gas humidification, which in particular are applied in mechanical ventilation.

Kinocilium

A kinocilium is a special type of cilium on the apex of hair cells located in the sensory epithelium of the vertebrate inner ear.

Anatomy in Humans

Kinocilia on hair cells in the human ear are involved in the senses of movement and hearing. In the auditory system, each hair cell has a kinocilium that is important in the depolarization or hyperpolarization of the cell. In the vestibular system, movement causes displacement of the stereocilia and kinocilium, resulting in depolarization or hyperpolarization of the cell. The depolarization of the hair cells in both instances causes signal transduction via neurotransmitter release.

Auditory System

In the human ear, each hair cell has a single non-motile cilium-like structure known as the kinocilium. The movement of other stereocilia on the hair cell towards the kinocilium, as a result of [[endolymph [1]]] flow, will cause potassium channels on the cilia to open due to a pulling force that each stereocilia exerts on its neighbouring cell (usually from tallest to shortest in a bundle of stereocilia) and this leads to the depolarization of the hair cell.

This pattern of depolarization should not be confused with the more common depolarization which involves the influx of Na+ into the cell which K+ channels stay closed. Endolymph composition resembles that of the intracellular fluid (more K+ and less Na+) more closely compared to its counterpart, perilymph which resembles the extracellular fluid (more Na+ and less K+ compared to intracellular matrix). This depolarization will open voltage gated calcium channels.

The influx of calcium then triggers the cell to release vesicles containing excitatory neurotransmitters into a synapse. The post-synaptic neurite then sends an action potential to the spiral ganglion. Cochlear hair cells have kinocilia that degenerate following maturation of the stereociliary bundle. It is believed that the kinocilium is important in delineating the anatomical asymmetry of stereocilia or in making some mechanical connections with the hair cells in which they persist.

Vestibular Apparatus

Kinocilia are present in the crista ampullaris of the semicircular ducts and the sensory maculae of the utricle and saccule. One kinocilium is the longest cilium located on the hair cell next to 40-70 sterocilia. During movement of the body, the hair cell is depolarized when the sterocilia move toward the kinocilium. The depolarization of the hair cell causes neurotransmitter to be released and an increase in firing frequency of cranial nerve VIII. When the sterocilia tilt away from the kinocilium, the hair cell is hyperpolarized, decreasing the amount of neurotransmitter released, which decreases the firing frequency of cranial nerve VIII.

Anatomy in Fish and Frogs

The apical surface of a sensory fish hair cell usually has numerous stereocilia and a single, much longer kinocilium. Deflection of the stereocilia toward or away from the kinocilium causes an increase or decrease in the firing rate of the sensory neuron innervating the hair cell at its basal surface. Hair cells in fish and some frogs are used to detect water movements around their bodies. These hair cells are embedded in a jelly-like protrusion called cupula. The hair cells therefore can not be seen and do not appear on the surface of skin of fish and frogs.

Flagellum

A flagellum is a tail-like projection that protrudes from the cell body of certain prokaryotic and eukaryotic cells, and functions in locomotion. There are some notable differences between prokaryotic and eukaryotic flagella, such as protein composition, structure, and mechanism of propulsion. An example of a flagellated bacterium is the ulcer-causing *Helicobacter pylori*, which uses multiple flagella to propel itself through the mucus lining to reach the stomach epithelium. An example of a eukaryotic flagellated cell is the sperm cell, which uses its flagellum to propel itself through the female reproductive tract. Eukaryotic flagella are structurally identical to eukaryotic cilia, although distinctions are sometimes made according to function and/or length. The word *flagellum* is the Latin word for whip.

Types

Three types of flagella have so far been distinguished; bacterial, archaeal and eukaryotic. The main differences among these three types are summarized below:

- Bacterial flagella are helical filaments that rotate like screws. They provide two of several kinds of bacterial motility.
- Archaeal flagella are superficially similar to bacterial flagella, but are different in many details and considered non-homologous.
- Eukaryotic flagella - those of animal, plant, and protist cells - are complex cellular projections that lash back and forth.

Sometimes eukaryotic flagella are called *cilia* or *undulipodia* to emphasize their distinctiveness.

Bacterial

The bacterial flagellum is made up of the protein flagellin. Its shape is a 20 nanometer-thick hollow tube. It is helical and has a sharp bend just outside the outer membrane; this "hook" allows the helix to point directly away from the cell. A shaft runs between the hook and the basal body, passing through protein rings in the cell's membrane that act as bearings. Gram-positive organisms have 2 of these basal body rings, one in the peptidoglycan layer and one in the plasma membrane. Gram-negative organisms have 4 such rings: the L ring associates with the lipopolysaccharides, the P ring associates with peptidoglycan layer, the M ring is embedded in the plasma membrane, and the S ring is directly attached to the plasma membrane. The filament ends with a capping protein.

The bacterial flagellum is driven by a rotary engine (the *Mot complex*) made up of protein, located at the flagellum's anchor point on the inner cell membrane. The engine is powered by proton motive force, i.e., by the flow of protons (hydrogen ions) across the bacterial cell membrane due to a concentration gradient set up by the cell's metabolism (in *Vibrio* species there are two kinds of flagella, lateral and polar, and some are driven by a sodium ion pump rather than a proton pump). The rotor transports protons across the membrane, and is turned in the process. The rotor alone can operate at 6,000 to 17,000 rpm, but with the flagellar filament attached usually only reaches 200 to 1000 rpm. The direction of rotation can be switched almost instantaneously, caused by a slight change in the position of a protein, FliG, in the rotor.

The cylindrical shape of flagella is suited to locomotion of microscopic organisms; these organisms operate at a low Reynolds number, where the viscosity of the surrounding water is much more important than its mass or inertia.

Flagella do not rotate at a constant speed but instead can increase or decrease their rotational speed in relation to the strength of the proton motive force. Flagellar rotation can move bacteria through liquid media at speeds of up to 60 cell lengths/second (sec). Although this is only about 0.00017 km/h (0.00011 mph), when comparing this speed with that of higher organisms in terms of number of lengths moved per second, it is extremely fast. By comparison, the cheetah, the fastest land animal, can sprint at 110 km/h (68 mph), which is approximately 25 body lengths/sec. During flagellar assembly, components of the flagellum pass through the hollow cores of the basal body and the nascent filament. During assembly, protein components are added at the flagellar tip rather than at the base. *In vitro*, flagellar filaments assemble spontaneously in a solution containing purified flagellin as the sole protein.

The flagellar filament is the long helical screw that propels the bacterium when rotated by the motor, through the hook. In most bacteria that have been studied, including the Gram negative *Escherichia coli, Salmonella typhimurium, Caulobacter crescentus*, and *Vibrio alginolyticus*, the filament is made up of eleven protofilaments approximately parallel to the filament axis. Each protofilament is a series of tandem protein chains. However in *Campylobacter jejuni*, there are seven protofilaments.

The basal body has several traits in common with some types of secretory pores, such as the hollow rod-like "plug" in their centres extending out through the plasma membrane. Given the structural similarities between bacterial flagella and bacterial secretory systems, it is thought that bacterial flagella may have evolved from the type three secretion system; however, it is not known for certain whether these pores are derived from the bacterial flagella or the bacterial secretory system. Through use of their flagella, *E. coli* are able to move rapidly towards attractants and away from repellents. They do this by means of a biased random walk, with 'runs' and 'tumbles' brought about by rotating the flagellum counter-clockwise and clockwise respectively.

Flagella Arrangement Schemes

Different species of bacteria have different numbers and arrangements of flagella. Monotrichous bacteria have a single flagellum (e.g., *Vibrio cholerae*). Lophotrichous bacteria have multiple flagella located at the same spot on the bacteria's surfaces which act in concert to drive the bacteria in a single direction. In many cases, the bases of multiple flagella are surrounded by a specialized region of the cell membrane, the so-called *polar membrane*. Amphitrichous bacteria have a single flagellum on each of two opposite ends (only one flagellum operates at a time, allowing the bacteria to reverse course rapidly by switching which flagellum is active). Peritrichous bacteria have flagella projecting in all directions (e.g., *E. coli*).

In some bacteria, such as the larger forms of *Selenomonas*, the individual flagella are organized outside the cell body, helically twining about each other to form a thick structure called a "fascicle". Other bacteria, such as Spirochetes, have a specialized type of flagellum called an "axial filament" that is located in the periplasmic space, the rotation of which causes the entire bacterium to move forward in a corkscrew-like motion.

Counterclockwise rotation of monotrichous polar flagella pushes the cell forward with the flagella trailing behind, much like a corkscrew moving inside cork. Indeed water in the microscopic scale is highly viscous, very different from our daily experience of water. The flagella are left-handed helices, and bundle and rotate together only when rotating counterclockwise. When some of the rotors reverse direction, the flagella unwind and the cell starts "tumbling". It has also been suggested that even if all flagella would rotate clockwise, they will not form a bundle, due to geometrical as well as hydrodynamical reasons.

Such "tumbling" may happen occasionally, leading to the cell seemingly thrashing about in place, resulting in the reorientation of the cell. The clockwise rotation of a flagellum is suppressed by chemical compounds favourable to the cell (e.g. food), but the motor is highly adaptive to this. Therefore, when moving in a favourable direction, the concentration of the chemical attractant increases and "tumbles" are continually suppressed; however, when the cell's direction of motion is unfavourable (e.g., away from a chemical attractant), tumbles are no longer suppressed and occur much more often, with the chance that the cell will be thus reoriented in the correct direction.

In some *Vibrio* spp. (particularly *Vibrio parahemolyticus*) and related proteobacteria such as *Aeromonas*, two flagellar systems co-exist, using different sets of genes and different ion gradients for energy. The polar flagella are constitutively expressed and provide motility in bulk fluid, while the lateral flagella are expressed when the polar flagella meet too much resistance to turn. These provide swarming motility on surfaces or in viscous fluids.

Archaeal

The archaeal flagellum is superficially similar to the bacterial (or eubacterial) flagellum; in the 1980s they were thought to be homologous on the basis of gross morphology and behavior. Both flagella consist of filaments extending outside of the cell, and rotate to propel the cell. Archaeal flagella have a unique structure which lacks a central channel. Similar to bacterial type IV pilins, the component flagellins are made with class 3 signal peptides and they are processed by a type IV prepilin peptidase-like enzyme. The archaeal flagellins are typically modified by the addition of N-linked glycans which are necessary for proper assembly and/or function.

Discoveries in the 1990s revealed numerous detailed differences between the archaeal and bacterial flagella; these include:

- Bacterial flagella are motorized by a flow of H^+ ions (or occasionally Na^+ ions); archaeal flagella are almost certainly powered by ATP. The torque-generating motor that powers rotation of the archaeal flagellum has not been identified.

- While bacterial cells often have many flagellar filaments, each of which rotates independently, the archaeal flagellum is composed of a bundle of many filaments that rotate as a single assembly.
- Bacterial flagella grow by the addition of flagellin subunits at the tip; archaeal flagella grow by the addition of subunits to the base.
- Bacterial flagella are thicker than archaeal flagella, and the bacterial filament has a large enough hollow "tube" inside that the flagellin subunits can flow up the inside of the filament and get added at the tip; the archaeal flagellum is too thin to allow this.
- Many components of bacterial flagella share sequence similarity to components of the type III secretion systems, but the components of bacterial and archaeal flagella share no sequence similarity. Instead, some components of archaeal flagella share sequence and morphological similarity with components of type IV pili, which are assembled through the action of type II secretion systems (the nomenclature of pili and protein secretion systems is not consistent).

These differences could mean that the bacterial and archaeal flagella could be a classic case of biological analogy, or convergent evolution, rather than homology. However, in comparison to the decades of well-publicized study of bacterial flagella (e.g. by Berg), archaeal flagella have only recently begun to get serious scientific attention. Therefore, many assume erroneously that there is only one basic kind of prokaryotic flagellum, and that archaeal flagella are homologous to it. For example, Cavalier-Smith (2002) is aware of the differences between archaeal and bacterial flagellins, but retains the misconception that the basal bodies are homologous.

Eukaryotic

Along with cilia, flagella make up a group of organelles known as undulipodia.

Structure

A eukaryotic flagellum is a bundle of nine fused pairs of microtubule *doublets* surrounding two central single microtubules. The so-called *"9+2"* structure is characteristic of the core of the eukaryotic flagellum called an *axoneme*. At the base of a eukaryotic flagellum is a basal body, "blepharoplast" or kinetosome, which is the microtubule organizing centre (MTOC) for flagellar microtubules and is about 500 nanometers long. Basal bodies are structurally identical to centrioles. The flagellum is encased within the cell's plasma membrane, so that the interior of the flagellum is accessible to the cell's cytoplasm.

Mechanism

Each of the outer 9 doublet microtubules extends a pair of dynein arms (an "inner" and an "outer" arm) to the adjacent microtubule; these dynein arms are responsible for flagellar beating, as the force produced by the arms causes the microtubule doublets to slide against each other and the flagellum as a whole to bend.

These dynein arms produce force through ATP hydrolysis. The flagellar axoneme also contains radial spokes, polypeptide complexes extending from each of the outer 9 microtubule doublets towards the central pair, with the "head" of the spoke facing inwards. The radial spoke is thought to be involved in the regulation of flagellar motion, although its exact function and method of action are not yet understood.

Flagella vs Cilia

Though eukaryotic flagella and motile cilia are ultrastructurally identical, the beating pattern of the two organelles can be different. In the case of flagella (e.g. the tail of a sperm) the motion is propeller-like. In contrast, beating of motile cilia consists of coordinated back-and-forth cycling of many cilia on the cell surface. Thus, flagella serve for the propulsion of single cells (e.g. swimming of protozoa and spermatozoa), and motile cilia for the transport of fluids (e.g. transport of mucus by stationary ciliated cells in the trachea). However, cilia are also used for locomotion (through liquids) in organisms such as *Paramecium*.

Intraflagellar Transport

Intraflagellar transport (IFT), the process by which axonemal subunits, transmembrane receptors, and other proteins are moved up and down the length of the flagellum, is essential for proper functioning of the flagellum, in both motility and signal transduction.

Evolution of Flagella and Debate

Based on similarity in structure and partial similarity in amino acid sequence, it is generally accepted among scientists that the eukaryotic flagellum and cilium have evolved from the cytoskeleton, while the eubacterial flagellum has evolved either from the type III secretion system or from a more ancient secretion system from which the type III secretion system has evolved as well. The archaeal flagellum has probably evolved from the type IV pili. More details appear under Evolution of flagella.

In his 1996 book *Darwin's Black Box*, intelligent design proponent Michael Behe, under funding from the Templeton Foundation, cited the bacterial flagellum as an example of an irreducibly complex structure that could not have evolved through naturalistic means. Behe argued that the flagellum becomes useless if any one of its constituent parts is removed, and thus could not have arisen through numerous, successive, slight modifications; therefore, it is hopelessly improbable that the proteins making up the flagellar motor could have come together all at once, by chance. Mark Perakh explained that while

Behe popularized the idea, biologist Hermann J. Müller had already explored it (under the slightly different name of "interlocking complexity") and more than a decade before Behe's book the same idea was explored by A. Graham Cairns-Smith, but neither claimed that "irreducible complexity" was a "marker" of a supernatural design.

While Behe discussed the immune system and the blood clotting cascade in greater detail, the bacterial flagellum has become a "poster child" for intelligent design proponents and other creationists, as during the Kitzmiller vs Dover Area School District trial of 2005. It is one of two identified rotary structures found in nature (the other being ATP synthase) and it is billions of years older than Behe's other two examples, which exist in many homologous forms, simplifying the explanation of their origin. Evolutionary pathways supported by the Theory of Natural Selection and Evolution have since been identified for the bacterial flagellum; thus, undermining Behe's argument. In addition, the Type three secretion system, a molecular syringe which bacteria use to inject toxins into other cells, appears to be a simplified sub-set of the bacterial flagellum's components, meaning that it is much less likely to be irreducibly complex in the way that the bacterial flagellum could have in fact evolved from the type three secretion system. Exaptation explains how systems with multiple parts can evolve through natural means.

Ciliopathy

A ciliopathy is a genetic disorder of the cellular cilia or the cilia anchoring structures, the basal bodies.

Although non-motile or primary cilia were first described in 1898, "cell biologists largely ignored them." But "microscopists continued to document their presence in the cells of most vertebrate organisms." The "primary cilium was long considered—with few exceptions—to be a largely useless evolutionary vestige, a vestigial organelle. Recent research has revealed an initial understanding that cilia are essential to many of the body's organs". Many mammalian eukaryotic cells are ciliated with *primary cilia*. These primary cilia play important roles in chemosensation, mechanosensation, and thermosensation. Cilia may thus be "viewed as sensory cellular antennae that coordinate a large number of cellular signalling pathways, sometimes coupling the signalling to ciliary motility or alternatively to cell division and differentiation."

Recent advances in mammalian genetic research have facilitated the elucidation of a molecular basis for a number of dysfunctional mechanisms in both motile and primary cilia structures of the cell. "Numerous critical developmental signalling pathways" essential to cellular development have been discovered. These are principally but not exclusively found in the *non-motile* or primary cilia. A number of common observable characteristics of mammalian genetic disorders and diseases are caused by ciliary dysgenesis and dysfunction. Once identified, these characteristics thus describe a set of hallmarks of a ciliopathy. Cilia have recently been implicated in a wide variety of human genetic diseases by "the discovery that numerous proteins involved in mammalian disease localize to the basal bodies and cilia." For example, in just a single area of human disease physiology, cystic renal disease, cilia-related genes and proteins have been identified to have causal effect in polycystic kidney disease, nephronophthisis, Senior-Loken syndrome type 5, orofaciodigital syndrome type 1 and Bardet-Biedl syndrome.

The Mechanism of Ciliary Function

"In effect, the [motile cilium] is a nanomachine composed of perhaps over 600 proteins in molecular complexes, many of which also function independently as nanomachines." Cilia "function as mechano- or chemosensors and as a cellular global positioning system to detect changes in the surrounding environment." For example, ciliary signalling plays a role in the initiation of cellular replacement after cell damage. In addition to this sensory role mediating specific signalling cues, cilia play "a secretory role in which a soluble protein is released to have an effect downstream of the fluid flow" in epithelial cells, and can of course mediate fluid flow directly in the case of motile cilia. Primary cilia in the retina play a role in transferring nourishment to the non-vascularized rod and cone cells from the vascularized cells several microns behind the surface of the retina.

Similar Genes Can Result in a Range of Different Diseases

"Just as different genes can contribute to similar diseases, so the same genes and families of genes can play a part in a range of different diseases." For example, in just two of the diseases caused by malfunctioning cilia, Meckel-Gruber syndrome and Bardet-Biedl syndrome, patients who carry mutations in genes associated with both diseases "have unique symptoms that are not seen in either condition alone." The genes linked to the two different conditions "interact with each other during development." Systems biologists are endeavoring to define functional modules containing multiple genes and then look at disorders whose phenotypes fit into such modules.

A particular phenotype can overlap "considerably with several conditions (ciliopathies) in which primary cilia are also implicated in pathogenicity. One emerging aspect is the wide spectrum of ciliopathy gene mutations found within different diseases."

Ciliopathies

"The phenotypic parameters that define a ciliopathy may be used to both recognize the cellular basis of a number of genetic disorders and to facilitate the diagnosis and treatment of some diseases of unknown etiology".

Syndromes and Diseases with a Known Cilia-based Causal Mechanism

- Alstrom syndrome,
- Bardet-Biedl syndrome
- early embryonic death (some cases)
- hydrocephalus (some cases)
- Joubert syndrome
- Meckel-Gruber syndrome
- nephronophthisis
- orofaciodigital syndrome type 1
- polycystic kidney disease
- polycystic liver disease
- primary ciliary dyskinesia
- retinal degeneration (some forms)
- Senior-Loken syndrome.

Implied or Suspected Ciliopathies

- agenesis of the corpus callosum
- anencephaly
- breathing abnomalities
- cerebellar vermis hypoplasia
- Dandy-Walker malformation
- diabetes
- Ellis-van Creveld syndrome
- exencephaly
- eye movement abnormalities
- liver disease
- hypoplasia of the corpus callosum
- hypotonia
- reproductive sterility
- Jeune asphyxiating thoracic dystrophy
- Juvenile myoclonic epilepsy (JME)
- Kartagener syndrome
- Marden-Walker syndrome
- "mental retardation/developmental delay" or "other cognitive defects"
- obesity

- polydactyly
- posterior encephalocele
- respiratory dysfunction
- "recurrent respiratory infections"
- renal cystic disease
- retinitis pigmentosa (some forms)
- sensorineural deafness
- situs inversus/Isomerism
- spina bifida.

Clinical Symptoms and Ciliac Roles

A wide variety of symptoms are potential clinical features of ciliopathy.

- Chemosensation abnormalities,, typically via ciliated epithelial cellular dysfunction.
- Defective thermosensation or mechanosensation, often via ciliated epithelial cellular dysfunction.
- Cellular motility dysfunction
- Issues with displacement of extracellular fluid
- Paracrine signal transduction abnormalities.

In organisms of normal health, cilia are critical for:

- organism development
- homeostasis
- reproduction.

Centrioles and Basal Bodies

A centriole is a barrel-shaped cell structure found in most animal eukaryotic cells, though absent in higher plants and most fungi. The walls of each centriole are usually composed of nine triplets of microtubules (protein of the cytoskeleton). Deviations from this structure include *Drosophila melanogaster* embryos, with nine doublets, and *Caenorhabditis elegans* sperm cells and early embryos, with nine singlets. An associated pair of centrioles, arranged perpendicularly and surrounded by an amorphous mass of dense material (the pericentriolar material) constitutes the compound structure known as the centrosome.

Cell Division

Centrioles are involved in the organization of the mitotic spindle and in the completion of cytokinesis. Centrioles were previously thought to be required for the formation of a mitotic spindle in animal cells. However, more recent experiments have demonstrated that cells whose centrioles have been removed via laser ablation can still progress through the G_1 stage of interphase before centrioles can be synthesized later in a de novo fashion. Additionally, mutant flies lacking centrioles can develop almost normally, although the adult flies lack flagella and cilia, a lack that underscores the requirement of centrioles for the formation of these organelles.

Cellular Organization

Centrioles are a very important part of centrosomes, which are involved in organizing microtubules in the cytoplasm. The position of the centriole determines the position of the nucleus and plays a crucial role in the spatial arrangement of the cell.

Ciliogenesis

In organisms with flagella and cilia, the position of these organelles is determined by the mother centriole, which becomes the basal body. An inability of cells to use centrioles to make functional cilia and flagella has been linked to a number of genetic and developmental diseases. In particular, the inability of centrioles to properly migrate prior to ciliary assembly has recently been linked to Meckel-Gruber syndrome.

Animal Development

Proper orientation of cilia via centriole positioning toward the posterior of embryonic node cells is critical for establishing left–right asymmetry during mammalian development.

Centriole Duplication

Cells in G_0 and G_1 usually contain two complete centrioles. The older of the two centrioles in a pair is termed the *mother centriole*, whereas the younger is termed the *daughter centriole*. During the cell division cycle, a new centriole grows from the side of each of the existing "mother" centrioles. After centriole duplication, the two pairs of centrioles remain attached to each other in an orthogonal configuration until mitosis, when the mother and daughter centrioles separate in a manner dependent upon the enzyme separase.

The two centrioles in the centrosome are connected to each other by unidentified proteins. The mother centriole has radiating appendages at the distal end of its long axis and is attached to the daughter centriole at the other proximal end. Each daughter cell formed after cell division will inherit one of these pairs (one older and one newer centriole). Duplication of centrioles starts at the time of the G_1/S transition and ends before the onset of mitosis.

A basal body (also called a basal granule or kinetosome) is an organelle formed from a centriole, and a short cylindrical array of microtubules. It is found at the base of a eukaryotic undulipodium (cilium or flagellum) and serves as a nucleation site for the growth of the axoneme microtubules. Centrioles, from which basal bodies are derived, act as anchoring sites for proteins that in turn anchor microtubules within centrosomes, one type of microtubule organizing centre (MTOC). These microtubules provide structure and facilitate movement of vesicles and organelles within many eukaryotic cells. Basal bodies, however, are specifically the bases for cilia and flagella that extend out of the cell.

Basal bodies are derived from centrioles through a largely mysterious process. They are structurally the same, each containing a microtubule triplet 9*3 helocoidal configuration forming a hollow cylinder.

Regulation of basal body production and spatial orientation is a function of the nucleotide-binding domain of ã-tubulin (Shang et al., 2005).

Flagellum(flagella) are basically attached to the cell membrane from a basal granule.

7

The Nucleus

In cell biology, the nucleus (pl. *nuclei*; from Latin *nucleus* or *nuculeus*, meaning kernel) is a membrane-enclosed organelle found in eukaryotic cells. It contains most of the cell's genetic material, organized as multiple long linear DNA molecules in complex with a large variety of proteins, such as histones, to form chromosomes. The genes within these chromosomes are the cell's nuclear genome. The function of the nucleus is to maintain the integrity of these genes and to control the activities of the cell by regulating gene expression — the nucleus is therefore the control centre of the cell. The main structures making up the nucleus are the nuclear envelope, a double membrane that encloses the entire organelle and separates its contents from the cellular cytoplasm, and the nuclear lamina, a meshwork within the nucleus that adds mechanical support, much like the cytoskeleton supports the cell as a whole. Because the nuclear membrane is impermeable to most molecules, nuclear pores are required to allow movement of molecules across the envelope. These pores cross both of the membranes, providing a channel that allows free movement of small molecules and ions. The movement of larger molecules such as proteins is carefully controlled, and requires active transport regulated by carrier proteins. Nuclear transport is crucial to cell function, as movement through the pores is required for both gene expression and chromosomal maintenance.

Although the interior of the nucleus does not contain any membrane-bound subcompartments, its contents are not uniform, and a number of *subnuclear bodies* exist, made up of unique proteins, RNA molecules, and particular parts of the chromosomes. The best known of these is the nucleolus, which is mainly involved in the assembly of ribosomes. After being produced in the nucleolus, ribosomes are exported to the cytoplasm where they translate mRNA.

History

The nucleus was the first organelle to be discovered. The probably oldest preserved drawing dates back to the early microscopist Antonie van Leeuwenhoek (1632 – 1723). He observed a "Lumen", the nucleus, in the red blood cells of salmon. Unlike mammalian red blood cells, those of other vertebrates still possess nuclei. The nucleus was also described by Franz Bauer in 1804 and in more detail in 1831 by Scottish botanist Robert Brown in a talk at the Linnean Society of London. Brown was studying orchids microscopically when he observed an opaque area, which he called the areola or nucleus, in the cells of the flower's outer layer. He did not suggest a potential function. In 1838 Matthias Schleiden proposed that the nucleus plays a role in generating cells, thus he introduced the name "Cytoblast" (cell builder). He believed that he had observed new cells assembling around "cytoblasts". Franz Meyen was a strong opponent of this view having already described cells multiplying by division and believing that many cells would have no nuclei. The idea that cells can be generated de novo, by the "cytoblast" or otherwise, contradicted work by Robert Remak (1852) and Rudolf Virchow (1855) who decisively propagated the new paradigm that cells are generated solely by cells ("Omnis cellula e cellula"). The function of the nucleus remained unclear.

Between 1876 and 1878 Oscar Hertwig published several studies on the fertilization of sea urchin eggs, showing that the nucleus of the sperm enters the oocyte and fuses with its nucleus. This was the first time it was suggested that an individual develops from a (single) nucleated cell. This was in contradiction to Ernst Haeckel's theory that the complete phylogeny of a species would be repeated during embryonic development, including generation of the first nucleated cell from a "Monerula", a structureless mass of primordial mucus ("Urschleim"). Therefore, the necessity of the sperm nucleus for fertilization was discussed for quite some time. However, Hertwig confirmed his observation in other animal groups, e.g. amphibians and molluscs. Eduard Strasburger produced the same results for plants (1884). This paved the way to assign the nucleus an important role in heredity. In 1873 August Weismann postulated the equivalence of the maternal and paternal germ *cells* for heredity. The function of the nucleus as carrier of genetic information became clear only later, after mitosis was discovered and the Mendelian rules were rediscovered at the beginning of the 20th century; the chromosome theory of heredity was developed.

Structures

The nucleus is the largest cellular organelle in animals. In mammalian cells, the average diameter of the nucleus is approximately 6 micrometers (ìm), which occupies about 10% of the total cell volume. The viscous liquid within it is called nucleoplasm, and is similar in composition to the cytosol found outside the nucleus. It appears as a dense, roughly spherical organelle.

Nuclear Envelope and Pores

The nuclear envelope otherwise known as nuclear membrane consists of two cellular membranes, an inner and an outer membrane, arranged parallel to one another and separated by 10 to 50 nanometers (nm). The nuclear envelope completely encloses the nucleus and separates the cell's genetic material from the surrounding cytoplasm, serving as a barrier to prevent macromolecules from diffusing freely between the nucleoplasm and the cytoplasm. The outer nuclear membrane is continuous with the membrane of the rough endoplasmic reticulum (RER), and is similarly studded with ribosomes. The space between the membranes is called the perinuclear space and is continuous with the RER lumen.

Nuclear pores, which provide aqueous channels through the envelope, are composed of multiple proteins, collectively referred to as nucleoporins. The pores are about 125 million daltons in molecular weight and consist of around 50 (in yeast) to 100 proteins (in vertebrates). The pores are 100 nm in total diameter; however, the gap through which molecules freely diffuse is only about 9 nm wide, due to the presence of regulatory systems within the centre of the pore. This size allows the free passage of small water-soluble molecules while preventing larger molecules, such as nucleic acids and larger proteins, from inappropriately entering or exiting the nucleus. These large molecules must be actively transported into the nucleus instead. The nucleus of a typical mammalian cell will have about 3000 to 4000 pores throughout its envelope, each of which contains a donut-shaped, eightfold-symmetric ring-shaped structure at a position where the inner and outer membranes fuse. Attached to the ring is a structure called the *nuclear basket* that extends into the nucleoplasm, and a series of filamentous extensions that reach into the cytoplasm. Both structures serve to mediate binding to nuclear transport proteins.

Most proteins, ribosomal subunits, and some RNAs are transported through the pore complexes in a process mediated by a family of transport factors known as karyopherins. Those karyopherins that mediate movement into the nucleus are also called importins, while those that mediate movement out of the nucleus are called exportins. Most karyopherins interact directly with their cargo, although some use adaptor proteins. Steroid hormones such as cortisol and aldosterone, as well as other small lipid-soluble molecules involved in intercellular signalling can diffuse through the cell membrane and into the cytoplasm, where they bind nuclear receptor proteins that are trafficked into the nucleus. There they serve as transcription factors when bound to their ligand; in the absence of ligand many such receptors function as histone deacetylases that repress gene expression.

Nuclear Lamina

In animal cells, two networks of intermediate filaments provide the nucleus with mechanical support: the nuclear lamina forms an organized meshwork on the internal face of the envelope, while less organized support is provided on the cytosolic face of the envelope. Both systems provide structural support for the nuclear envelope and anchoring sites for chromosomes and nuclear pores.

The nuclear lamina is mostly composed of lamin proteins. Like all proteins, lamins are synthesized in the cytoplasm and later transported into the nucleus interior, where they are assembled before being incorporated into the existing network of nuclear lamina. Lamins are also found inside the nucleoplasm where they form another regular structure, known as the *nucleoplasmic veil*, that is visible using fluorescence microscopy. The actual function of the veil is not clear, although it is excluded from the nucleolus and is present during interphase. The lamin structures that make up the veil bind chromatin and disrupting their structure inhibits transcription of protein-coding genes.

Like the components of other intermediate filaments, the lamin monomer contains an alpha-helical domain used by two monomers to coil around each other, forming a dimer structure called a coiled coil. Two of these dimer structures then join side by side, in an antiparallel arrangement, to form a tetramer called a *protofilament*. Eight of these protofilaments form a lateral arrangement that is twisted to form a ropelike *filament*. These filaments can be assembled or disassembled in a dynamic manner, meaning that changes in the length of the filament depend on the competing rates of filament addition and removal. Mutations in lamin genes leading to defects in filament assembly are known as *laminopathies*. The most notable laminopathy is the family of diseases known as progeria, which causes the appearance of premature aging in its sufferers. The exact mechanism by which the associated biochemical changes give rise to the aged phenotype is not well understood.

Chromosomes

The cell nucleus contains the majority of the cell's genetic material, in the form of multiple linear DNA molecules organized into structures called chromosomes. During most of the cell cycle these are organized in a DNA-protein complex known as chromatin, and during cell division the chromatin can be seen to form the well defined chromosomes familiar from a karyotype.

A small fraction of the cell's genes are located instead in the mitochondria. There are two types of chromatin. Euchromatin is the less compact DNA form, and contains genes that are frequently expressed by the cell. The other type, heterochromatin, is the more compact form, and contains DNA that are infrequently transcribed. This structure is further categorized into *facultative* heterochromatin, consisting of genes that are organized as heterochromatin only in certain cell types or at certain stages of development, and *constitutive* heterochromatin that consists of chromosome structural components such as telomeres and centromeres. During interphase the chromatin organizes itself into discrete individual patches, called *chromosome territories*. Active genes, which are generally found in the euchromatic region of the chromosome, tend to be located towards the chromosome's territory boundary.

Antibodies to certain types of chromatin organization, particularly nucleosomes, have been associated with a number of autoimmune diseases, such as systemic lupus erythematosus. These are known as anti-nuclear antibodies (ANA) and have also been observed in concert with multiple sclerosis as part of general immune system dysfunction. As in the case of progeria, the role played by the antibodies in inducing the symptoms of autoimmune diseases is not obvious.

Nucleolus

The nucleolus is a discrete densely stained structure found in the nucleus. It is not surrounded by a membrane, and is sometimes called a *suborganelle*. It forms around tandem repeats of rDNA, DNA coding for ribosomal RNA (rRNA). These regions are called nucleolar organizer regions (NOR). The main roles of the nucleolus are to synthesize rRNA and assemble ribosomes. The structural cohesion of the nucleolus depends on its activity, as ribosomal assembly in the nucleolus results in the transient association of nucleolar components, facilitating further ribosomal assembly, and hence further association. This model is supported by observations that inactivation of rDNA results in intermingling of nucleolar structures.

The first step in ribosomal assembly is transcription of the rDNA, by a protein called RNA polymerase I, forming a large pre-rRNA precursor. This is cleaved into the subunits 5.8S, 18S, and 28S rRNA. The transcription, post-transcriptional processing, and assembly of rRNA occurs in the nucleolus, aided by small nucleolar RNA (snoRNA) molecules, some of which are derived from spliced introns from messenger RNAs encoding genes related to ribosomal function. The assembled ribosomal subunits are the largest structures passed through the nuclear pores.

When observed under the electron microscope, the nucleolus can be seen to consist of three distinguishable regions: the innermost *fibrillar centres* (FCs), surrounded by the *dense fibrillar component* (DFC), which in turn is bordered by the *granular component* (GC). Transcription of the rDNA occurs either in the FC or at the FC-DFC boundary, and therefore when rDNA transcription in the cell is increased more FCs are detected. Most of the cleavage and modification of rRNAs occurs in the DFC, while the latter steps involving protein assembly onto the ribosomal subunits occur in the GC.

Other Subnuclear Bodies

Table: *Subnuclear structure sizes.*

Structure name	*Structure diameter*
Cajal bodies	0.2–2.0 μm
PIKA	5 μmPML
bodies	0.2–1.0 μm
Paraspeckles	0.2–1.0 μm
Speckles	20–25 nm

Besides the nucleolus, the nucleus contains a number of other non-membrane delineated bodies. These include Cajal bodies, Gemini of coiled bodies, polymorphic interphase karyosomal association (PIKA), promyelocytic leukaemia (PML) bodies, paraspeckles and splicing speckles. Although little is known about a number of these domains, they are significant in that they show that the nucleoplasm is not uniform mixture, but rather contains organized functional subdomains.

Other subnuclear structures appear as part of abnormal disease processes. For example, the presence of small intranuclear rods have been reported in some cases of nemaline myopathy. This condition typically results from mutations in actin, and the rods themselves consist of mutant actin as well as other cytoskeletal proteins.

Cajal Bodies and Gems

A nucleus typically contains between 1 and 10 compact structures called Cajal bodies or coiled bodies (CB), whose diameter measures between 0.2 μm and 2.0 μm depending on the cell type and species. When seen under an electron microscope, they resemble balls of tangled thread and are dense foci of distribution for the protein coilin. CBs are involved in a number of different roles relating to RNA processing, specifically small nucleolar RNA (snoRNA) and small nuclear RNA (snRNA) maturation, and histone mRNA modification.

Similar to Cajal bodies are Gemini of coiled bodies, or gems, whose name is derived from the Gemini constellation in reference to their close “twin” relationship with CBs. Gems are similar in size and shape to CBs, and in fact are virtually indistinguishable under the microscope. Unlike CBs, gems do not contain small nuclear ribonucleoproteins (snRNPs), but do contain a protein called *survivor of motor neurons* (SMN) whose function relates to snRNP biogenesis. Gems are believed to assist CBs in snRNP biogenesis, though it has also been suggested from microscopy evidence that CBs and gems are different manifestations of the same structure.

PIKA and PTF Domains

PIKA domains, or polymorphic interphase karyosomal associations, were first described in microscopy studies in 1991. Their function was and remains unclear, though they were not thought to be associated with active DNA replication, transcription, or RNA processing. They have been found to often associate with discrete domains defined by dense localization of the transcription factor PTF, which promotes transcription of snRNA.

PML Bodies

Promyelocytic leukaemia bodies (PML bodies) are spherical bodies found scattered throughout the nucleoplasm, measuring around 0.2–1.0 μm. They are known by a number of other names, including nuclear domain 10 (ND10), Kremer bodies, and PML oncogenic domains. They are often seen in the nucleus in association with Cajal bodies and cleavage bodies. It has been suggested that they play a role in regulating transcription.

Paraspeckles

Discovered by Fox et al. in 2002, paraspeckles are irregularly shaped compartments in the nucleus' interchromatin space. First documented in HeLa cells, where there are generally 10–30 per nucleus, paraspeckles are now known to also exist in all human primary cells, transformed cell lines and tissue sections. Their name is derived from their distribution in the nucleus; the "para" is short for parallel and the "speckles" refers to the splicing speckles to which they are always in close proximity.

Paraspeckles are dynamic structures that are altered in response to changes in cellular metabolic activity. They are transcription dependent and in the absence of RNA Pol II transcription, the paraspeckle disappears and all of its associated protein components (PSP1, p54nrb, PSP2, CFI(m)68 and PSF) form a crescent shaped perinucleolar cap in the nucleolus. This phenomenon is demonstrated during the cell cycle. In the cell cycle, paraspeckles are present during interphase and during all of mitosis except for telophase. During telophase, when the two daughter nuclei are formed, there is no RNA Pol II transcription so the protein components instead form a perinucleolar cap.

Splicing Speckles

Sometimes referred to as *interchromatin granule clusters* or as *splicing-factor compartments*, speckles are rich in splicing snRNPs and other splicing proteins necessary for pre-mRNA processing. Because of a cell's changing requirements, the composition and location of these bodies changes according to mRNA transcription and regulation via phosphorylation of specific proteins.

Function

The main function of the cell nucleus is to control gene expression and mediate the replication of DNA during the cell cycle. The nucleus provides a site for genetic transcription that is segregated from the location of translation in the cytoplasm, allowing levels of gene regulation that are not available to prokaryotes.

Cell Compartmentalization

The nuclear envelope allows the nucleus to control its contents, and separate them from the rest of the cytoplasm where necessary. This is important for controlling processes on either side of the nuclear membrane. In some cases where a cytoplasmic process needs to be restricted, a key participant is removed to the nucleus, where it interacts with transcription factors to downregulate the production of certain enzymes in the pathway. This regulatory mechanism occurs in the case of glycolysis, a cellular pathway for breaking down glucose to produce energy. Hexokinase is an enzyme responsible for the first the step of glycolysis, forming glucose-6-phosphate from glucose. At high concentrations of fructose-6-phosphate, a molecule made later from glucose-6-phosphate, a regulator protein removes hexokinase to the nucleus, where it forms a transcriptional repressor complex with nuclear proteins to reduce the expression of genes involved in glycolysis.

In order to control which genes are being transcribed, the cell separates some transcription factor proteins responsible for regulating gene expression from physical access to the DNA until they are activated by other signalling pathways. This prevents even low levels of inappropriate gene expression. For example in the case of NF-κB-controlled genes, which are involved in most inflammatory responses, transcription is induced in response to a signal pathway such as that initiated by the signalling molecule TNF-α, binds to a cell membrane receptor, resulting in the recruitment of signalling proteins, and eventually activating the transcription factor NF-κB. A nuclear localization signal on the NF-κB protein allows it to be transported through the nuclear pore and into the nucleus, where it stimulates the transcription of the target genes.

The compartmentalization allows the cell to prevent translation of unspliced mRNA. Eukaryotic mRNA contains introns that must be removed before being translated to produce functional proteins. The splicing is done inside the nucleus before the mRNA can be accessed by ribosomes for translation. Without the nucleus ribosomes would translate newly transcribed (unprocessed) mRNA resulting in misformed and nonfunctional proteins.

Gene Expression

Gene expression is the process by which information from a gene is used in the synthesis of a functional gene product. These products are often proteins, but in non-protein coding genes such as rRNA genes or tRNA genes, the product is a functional RNA. The process of gene expression is used by all known life - eukaryotes (including multicellular organisms), prokaryotes (bacteria and archaea) and viruses - to generate the macromolecular machinery for life. Several steps in the gene expression process may be modulated, including the transcription, RNA splicing, translation, and post-translational modification of a protein. Gene regulation gives the cell control over structure and function, and is the basis for cellular differentiation, morphogenesis and the versatility and adaptability of any organism. Gene regulation may also serve as a substrate for evolutionary change, since control of the timing, location, and amount of gene expression can have a profound effect on the functions (actions) of the gene in a cell or in a multicellular organism. In genetics, gene expression is the most fundamental level at which genotype gives rise to the phenotype. The genetic code is "interpreted" by gene expression, and the properties of the expression products give rise to the organism's phenotype.

Mechanism

Transcription

The gene itself is typically a long stretch of DNA. It is a blueprint for the production of RNA. The production of RNA copies of the DNA is called transcription, and is performed by RNA polymerase, which adds one RNA nucleotide at a time to a growing RNA strand. This RNA is complementary to the template 3' '! 5' DNA strand, which is itself complementary to the coding 5' '! 3' DNA strand. Therefore, the resulting 5' '! 3' RNA strand is identical to the coding DNA strand with the exception that thymines are replaced with uracils in the RNA. A coding DNA strand reading "ATG" is transcribed as "AUG" in RNA.

RNA Processing

Transcription of protein encoding genes creates a primary transcript of RNA at the place where the gene was located. This transcript can be altered before being translated, this is particularly common in eukaryotes. The most common RNA processing is splicing to remove introns. Introns are RNA segments which are not found in the mature RNA, although they can function as precursors, e.g. for snoRNAs, which are RNAs that direct modification of nucleotides in other RNAs. Introns are common in eukaryotic genes but rare in prokaryotes. RNA processing, also known as post-transcriptional modification, can start during transcription, as is the case for splicing, where the spliceosome removes introns from newly formed RNA. Extensive RNA processing may be an evolutionary advantage made possible by the nucleus of eukaryotes. In prokaryotes transcription and translation happen together whilst in eukaryotes the nuclear membrane separates the two processes giving time for RNA processing to occur.

Non-coding RNA Maturation

In most organisms non-coding genes (ncRNA) are transcribed as precursors which undergo further processing. In the case of ribosomal RNAs (rRNA), they are often transcribed as a pre-rRNA which contains one or more rRNAs, the pre-rRNA is cleaved and modified (22 -O-methylation and pseudouridine formation) at a specific sites by approximately 150 different small nucleolus-restricted RNA species, called small nucleolar RNAs(snoRNAs), which like snRNAs, snoRNAs associate with proteins, forming snoRNPs. In eukaryotes, in particular a snoRNP, called RNase MRP cleaves the 45S pre-rRNA into the 28S, 5.8S, and 18S rRNAs. The rRNA and RNA processing factors are form large aggregates called the nucleolus.

In the case of transfer RNA (tRNA), for example, the 5' sequence is removed by RNase P, whereas the 3' end is removed by the tRNase Z enzyme. In the case of micro RNA (miRNA), miRNAs are first transcribed as primary transcripts or pri-miRNA with a cap and poly-A tail and processed to short, 70-nucleotide stem-loop structures known as pre-miRNA in the cell nucleus by the enzymes Drosha and Pasha, after being exported, it is then processed to mature miRNAs in the cytoplasm by interaction with the endonuclease Dicer, which also initiates the formation of the RNA-induced silencing complex (RISC), composed of the Argonaute protein.

RNA Export

In eukaryotes most mature RNA must be exported to the cytoplasm from the nucleus. While some RNAs function in the nucleus, many RNAs are transported through the nuclear pores and into the cytosol. Notably this includes all RNA types involved in protein synthesis. In some cases RNAs are additionally transported to a specific part of the cytoplasm, such as a synapse; they are then towed by motor proteins that bind through linker proteins to specific sequences (called "zipcodes") on the RNA.

Translation

For some RNA (non-coding RNA) the mature RNA is the finished gene product. In the case of messenger RNA (mRNA) the RNA is an information carrier coding for the synthesis of one or more proteins. mRNA carrying a single protein sequence (common in eukaryotes) is monocistronic whilst mRNA carrying multiple protein sequences (common in prokaryotes) is known as polycistronic.

Each triplet of nucleotides of the coding regions of a messenger RNA corresponds to a binding site for a transfer RNA. Transfer RNAs carry amino acids, and these are chained together by the ribosome. The ribosome helps transfer RNA to bind to messenger RNA and takes the amino acid from each transfer RNA and makes a structure-less protein out of it. In prokaryotes translation generally occurs at the point of transcription, often using a messenger RNA which is still in the process of being created. In eukaryotes translation can occur in a variety of regions of the cell depending on where the protein being written is supposed to be. Major locations are the cytoplasm for soluble cytoplasmic proteins and the endoplasmic reticulum for proteins which are for export from the cell or insertion into a cell membrane. Proteins which are supposed to be expressed at the endoplasmic reticulum are recognised part-way through the translation process. This is governed by the signal recognition particle - a protein which binds to the ribosome and directs it to the endoplasmic reticulum when it finds a signal sequence on the growing (nascent) amino acid chain.

Folding

Enzymes called chaperones assist the newly formed protein to attain (fold into) the 3-dimensional structure it needs to function. Similarly, RNA chaperones help RNAs attain their functional shapes. Assisting protein folding is one of the main roles of the endoplasmic reticulum in eukaryotes.

Protein Transport

Many proteins are destined for other parts of the cell than the cytosol and a wide range of signalling sequences are used to direct proteins to where they are supposed to be. In prokaryotes this is normally a simple process due to limited compartmentalisation of the cell. However in eukaryotes there is a great variety of different targeting processes to ensure the protein arrives at the correct organelle. Not all proteins remain within the cell and many are exported, for example digestive enzymes, hormones and extracellular matrix proteins. In eukaryotes the export pathway is well developed and the main mechanism for the export of these proteins is translocation to the endoplasmatic reticulum, followed by transport via the Golgi apparatus.

Measurement

Measuring gene expression is an important part of many life sciences - the ability to quantify the level at which a particular gene is expressed within a cell, tissue or organism can give a huge amount of information. For example measuring gene expression can:

- Identify viral infection of a cell (viral protein expression)
- Determine an individual's susceptibility to cancer (oncogene expression)
- Find if a bacterium is resistant to penicillin (beta-lactamase expression).

Similarly the analysis of the location of expression protein is a powerful tool and this can be done on an organism or cellular scale. Investigation of localisation is particularly important for study of development in multicellular organisms and as an indicator of protein function in single cells. Ideally measurement of expression is done by detecting the final gene product (for many genes this is the protein) however it is often easier to detect one of the precursors, typically mRNA, and infer gene expression level.

mRNA Quantification

Levels of mRNA can be quantitatively measured by Northern blotting which gives size and sequence information about the mRNA molecules. A sample of RNA is separated on an agarose gel and hybridized to a radio-labeled RNA probe that is complementary to the target sequence. The radio-labeled RNA is then detected by an autoradiograph. The main problems with Northern blotting stem from the use of radioactive reagents (which make the procedure time consuming and potentially dangerous) and lower quality quantification than more modern methods (due to the fact that quantification is done by measuring band strength in an image of a gel). Northern blotting is, however, still widely used as the additional mRNA size information allows the discrimination of alternately spliced transcripts.

A more modern low-throughput approach for measuring mRNA abundance is reverse transcription quantitative polymerase chain reaction (RT-PCR followed with qPCR). RT-PCR first generates a DNA template from the mRNA by reverse transcription, which is called cDNA. This cDNA template is then used for qPCR where the change in fluorescence of a probe changes as the DNA amplification process progresses. With a carefully constructed standard curve qPCR can produce an absolute measurement such as number of copies of mRNA, typically in units of copies per nanolitre of homogenized tissue or copies per cell. qPCR is very sensitive (detection of a single mRNA molecule is possible), but can be expensive due to the fluorescent probes required.

Northern blots and RT-qPCR are good for detecting whether a single gene is being expressed, but it quickly becomes impractical if many genes within the sample are being studied. Using DNA microarrays transcript levels for many genes at once (expression profiling) can be measured. Recent advances in microarray technology allow for the quantification, on a single array, of transcript levels for every known gene in several organism's genomes, including humans.

Alternatively "tag based" technologies like Serial analysis of gene expression (SAGE), which can provide a relative measure of the cellular concentration of different messenger RNAs, can be used. The great advantage of tag-based methods is the "open architecture", allowing for the exact measurement of any transcript, with a known or unknown sequence.

Protein Quantification

For genes encoding proteins the expression level can be directly assessed by a number of means with some clear analogies to the techniques for mRNA quantification. The most commonly used method is to perform a Western blot against the protein of interest - this gives information on the size of the protein in addition to its identity. A sample (often cellular lysate) is separated on a polyacrylamide gel, transferred to a membrane and then probed with an antibody to the protein of interest. The antibody can either be conjugated to a fluorophore or to horseradish peroxidase for imaging and/or quantification. The gel-based nature of this assay makes quantification less accurate but it has the advantage of being able to identify later modifications to the protein, for example proteolysis or ubiquitination, from changes in size.

By replacing the gene with a new version fused a green fluorescent protein (or similar) marker expression may be directly quantified in live cells. This is done by imaging using a fluorescence microscope. It is very difficult to clone a GFP-fused protein into its native location in the genome without affecting expression levels so this method often cannot be used to measure endogenous gene expression.

It is, however, widely used to measure the expression of a gene artificially introduced into the cell, for example via an expression vector. It is important to note that by fusing a target protein to a fluorescent reporter the protein's behaviour, including its cellular localization and expression level, can be significantly changed. The enzyme-linked immunosorbent assay works by using antibodies immobilised on a microtiter plate to capture proteins of interest from samples added to the well. Using a detection antibody conjugated to an enzyme or fluorophore the quantity of bound protein can be accurately measured by fluorometric or colourimetric detection. The detection process is very similar to that of a Western blot, but by avoiding the gel steps more accurate quantification can be achieved.

Localization

Analysis of expression is not limited to only quantification; localization can also be determined. mRNA can be detected with a suitably labelled complementary mRNA strand and protein can be detected via labelled antibodies. The probed sample is then observed by microscopy to identify where the mRNA or protein is.

Regulation of Gene Expression

Regulation of gene expression refers to the control of the amount and timing of appearance of the functional product of a gene. Control of expression is vital to allow a cell to produce the gene products it needs when it needs them; in turn this gives cells the flexibility to adapt to a variable environment, external signals, damage to the cell, etc. Some simple examples of where gene expression is important are:

- Control of insulin expression so it gives a signal for blood glucose regulation
- X chromosome inactivation in female mammals to prevent an "overdose" of the genes it contains.
- Cyclin expression levels control progression through the eukaryotic cell cycle.

More generally gene regulation gives the cell control over all structure and function, and is the basis for cellular differentiation, morphogenesis and the versatility and adaptability of any organism. Any step of gene expression may be modulated, from the DNA-RNA transcription step to post-translational modification of a protein. The stability of the final gene product, whether it is RNA or protein, also contributes to the expression level of the gene - an unstable product results in a low expression level. In general gene expression is regulated through changes in the number and type of interactions between molecules that collectively influence transcription of DNA and translation of RNA.

Transcriptional Regulation

Regulation of transcription can be broken down into three main routes of influence; genetic (direct interaction of a control factor with the gene), modulation (interaction of a control factor with the transcription machinery) and epigenetic (non-sequence changes in DNA structure which influence transcription). Direct interaction with DNA is the simplest and most direct method a protein can change transcription levels and genes often have several protein binding sites around the coding region with the specific function of regulating transcription. There are many classes of regulatory DNA binding sites known as enhancers, insulators, repressors and silencers. The mechanisms for regulating transcription are very varied, from blocking key binding sites on the DNA for RNA polymerase to acting as an activator and promoting transcription by assisting RNA polymerase binding.

The activity of transcription factors is further modulated by intra-cellular signals causing protein post-translational modification including phosphorylated, acetylated, or glycosylated. These changes influence a transcription factor's ability to bind, directly or indirectly, to promoter DNA, to recruit RNA polymerase, or to favour elongation of a newly synthetized RNA molecule.

The nuclear membrane in eukaryotes allows further regulation of transcription factors by the duration of their presence in the nucleus which is regulated by reversible changes in their structure and by binding of other proteins. Environmental stimuli or endocrine signals may cause modification of regulatory proteins eliciting cascades of intracellular signals, which result in regulation of gene expression. More recently it has become apparent that there is a huge influence of non-DNA-sequence specific effects on translation. These effects are referred to as epigenetic and involve the higher order structure of DNA, non-sequence specific DNA binding proteins and chemical modification of DNA. In general epigenetic effects alter the accessibility of DNA to proteins and so modulate transcription.

DNA methylation is a widespread mechanism for epigenetic influence on gene expression and is seen in bacteria and eukaryotes and has roles in heritable transcription silencing and transcription regulation. In eukaryotes the structure of chromatin, controlled by the histone code, regulates access to DNA with significant impacts on the expression of genes in euchromatin and heterochromatin areas.

Post-transcriptional Regulation

In eukaryotes, where export of RNA is required before translation is possible, nuclear export is thought to provide additional control over gene expression. All transport in and out of the nucleus is via the nuclear pore and transport is controlled by a wide range of importin and exportin proteins. Expression of a gene coding for a protein is only possible if the messenger RNA carrying the code survives long enough to be translated. In a typical cell an RNA molecule is only stable if specifically protected from degradation.

RNA degradation has particular importance in regulation of expression in eukaryotic cells where mRNA has to travel significant distances before being translated. In eukaryotes RNA is stabilised by certain post-transcriptional modifications, particularly the 5' cap and poly-adenylated tail. Intentional degradation of mRNA is used not just as a defence mechanism from foreign RNA (normally from viruses) but also as a route of mRNA *destabilisation*. If an mRNA molecule has a complementary sequence to a small interfering RNA then it is targeted for destruction via the RNA interference pathway.

Translational Regulation

Direct regulation of translation is less prevalent than control of transcription or mRNA stability but is occasionally used. Inhibition of protein translation is a major target for toxins and antibiotics in order to kill a cell by overriding its normal gene expression control. Protein synthesis inhibitors include the antibiotic neomycin and the toxin ricin.

Protein Degradation

Once protein synthesis is complete the level of expression of that protein can be reduced by protein degradation. There are major protein degradation pathways in all prokaryotes and eukaryotes of which the proteasome is a common component. An unneeded or damaged protein is often labelled for degradation by addition of ubiquitin.

Expression System

An expression system is a system specifically designed for the production of a gene product of choice. This is normally a protein although may also be RNA, such as tRNA or a ribozyme. An expression system consists of a gene, normally encoded by DNA, and the molecular machinery required to transcribe the DNA into mRNA and translate the mRNA into protein using the reagents provided. In the broadest sense this includes every living cell but the term is more normally used to refer to expression as a laboratory tool. An expression system is therefore often artificial in some manner. Expression systems are, however, a fundamentally natural process. Viruses are an excellent example where they replicate by using the host cell as an expression system for the viral proteins and genome.

In Nature

In addition to these biological tools, certain naturally observed configurations of DNA (genes, promoters, enhancers, repressors) and the associated machinery itself are referred to as an expression system. This term is normally used in the case where a gene or set of genes is switched on under well defined conditions. For example the simple repressor 'switch' expression system in Lambda phage and the lac operator system in bacteria. Several natural expression systems are directly used or modified and used for artificial expression systems such as the Tet-on and Tet-off expression system.

Gene Networks and Expression

Genes have sometimes been regarded as nodes in a network, with inputs being proteins such as transcription factors, and outputs being the level of gene expression. The node itself performs a function, and the operation of these functions have been interpreted as performing a kind of information processing within cell and determine cellular behaviour.

Gene networks can also be constructed without formulating an explicit causal model. This is often the case when assembling networks from large expression data sets. Covariation and correlation of expression is computed across a large sample of cases and measurements (often transcriptome or proteome data). The source of variation can be either experimental or natural (observational). There are several ways to construct gene expression networks, but one common approach is to compute a matrix of all pair-wise correlations of expression across conditions, time points, or individuals and convert the matrix (after thresholding at some cut-off value) into a graphical representation in which nodes represent genes, transcripts, or proteins and edges connecting these nodes represent the strength of association.

Techniques and Tools

The following experimental techniques are used to measure gene expression and are listed in roughly chronological order, starting with the older, more established technologies. They are divided into two groups based on their degree of multiplexity.

- Low-to-mid-plex techniques:
  - o Reporter gene
  - o Northern blot
  - o Western blot
  - o Fluorescent in situ hybridization
  - o Reverse transcription PCR.
- Higher-plex techniques:
- o SAGE
- o DNA microarray
- o Tiling array
- o RNA-Seq.

Gene expression first involves transcription, in which DNA is used as a template to produce RNA. In the case of genes encoding proteins, that RNA produced from this process is messenger RNA (mRNA), which then needs to be translated by ribosomes to form a protein. As ribosomes are located outside the nucleus, mRNA produced needs to be exported.

Since the nucleus is the site of transcription, it also contains a variety of proteins which either directly mediate transcription or are involved in regulating the process.

These proteins include helicases that unwind the double-stranded DNA molecule to facilitate access to it, RNA polymerases that synthesize the growing RNA molecule, topoisomerases that change the amount of supercoiling in DNA, helping it wind and unwind, as well as a large variety of transcription factors that regulate expression.

Processing of Pre-mRNA

Newly synthesized mRNA molecules are known as primary transcripts or pre-mRNA. They must undergo post-transcriptional modification in the nucleus before being exported to the cytoplasm; mRNA that appears in the nucleus without these modifications is degraded rather than used for protein translation. The three main modifications are 5' capping, 3' polyadenylation, and RNA splicing. While in the nucleus, pre-mRNA is associated with a variety of proteins in complexes known as heterogeneous ribonucleoprotein particles (hnRNPs). Addition of the 5' cap occurs co-transcriptionally and is the first step in post-transcriptional modification. The 3' poly-adenine tail is only added after transcription is complete.

RNA splicing, carried out by a complex called the spliceosome, is the process by which introns, or regions of DNA that do not code for protein, are removed from the pre-mRNA and the remaining exons connected to re-form a single continuous molecule. This process normally occurs after 5' capping and 3' polyadenylation but can begin before synthesis is complete in transcripts with many exons. Many pre-mRNAs, including those encoding antibodies, can be spliced in multiple ways to produce different mature mRNAs that encode different protein sequences. This process is known as alternative splicing, and allows production of a large variety of proteins from a limited amount of DNA.

Post-transcriptional Modification

Post-transcriptional modification is a process in cell biology by which, in eukaryotic cells, primary transcript RNA is converted into mature RNA. A notable example is the conversion of precursor messenger RNA into mature messenger RNA (mRNA), which includes splicing and occurs prior to protein synthesis. This process is vital for the correct translation of the genomes of eukaryotes as the human primary RNA transcript that is produced as a result of transcription contains both exons, which are coding sections of the primary RNA transcript and introns, which are the non coding sections of the primary RNA transcript.

Dynamics and Regulation

Nuclear Transport

The entry and exit of large molecules from the nucleus is tightly controlled by the nuclear pore complexes. Although small molecules can enter the nucleus without regulation, macromolecules such as RNA and proteins require association karyopherins called importins to enter the nucleus and exportins to exit. “Cargo” proteins that must be translocated from the cytoplasm to the nucleus contain short amino acid sequences known as nuclear localization signals which are bound by importins, while those transported from the nucleus to the cytoplasm carry nuclear export signals bound by exportins. The ability of importins and exportins to transport their cargo is regulated by GTPases, enzymes that hydrolyze the molecule guanosine triphosphate to release energy. The key GTPase in nuclear transport is Ran, which can bind either GTP or GDP (guanosine diphosphate) depending on whether it is located in the nucleus or the cytoplasm. Whereas importins depend on RanGTP to dissociate from their cargo, exportins require RanGTP in order to bind to their cargo.

Nuclear import depends on the importin binding its cargo in the cytoplasm and carrying it through the nuclear pore into the nucleus. Inside the nucleus, RanGTP acts to separate the cargo from the importin, allowing the importin to exit the nucleus and be reused. Nuclear export is similar, as the exportin binds the cargo inside the nucleus in a process facilitated by RanGTP, exits through the nuclear pore, and separates from its cargo in the cytoplasm.

Specialized export proteins exist for translocation of mature mRNA and tRNA to the cytoplasm after post-transcriptional modification is complete. This quality-control mechanism is important due to the these molecules' central role in protein translation; mis-expression of a protein due to incomplete excision of exons or mis-incorporation of amino acids could have negative consequences for the cell; thus incompletely modified RNA that reaches the cytoplasm is degraded rather than used in translation.

Assembly and Disassembly

During its lifetime a nucleus may be broken down, either in the process of cell division or as a consequence of apoptosis, a regulated form of cell death. During these events, the structural components of the nucleus—the envelope and lamina—are systematically degraded. During the cell cycle the cell divides to form two cells. In order for this process to be possible, each of the new daughter cells must have a full set of genes, a process requiring replication of the chromosomes as well as segregation of the separate sets. This occurs by the replicated chromosomes, the sister chromatids, attaching to microtubules, which in turn are attached to different centrosomes. The sister chromatids can then be pulled to separate locations in the cell. In many cells the centrosome is located in the cytoplasm, outside the nucleus, the microtubules would be unable to attach to the chromatids in the presence of the nuclear envelope.

Therefore the early stages in the cell cycle, beginning in prophase and until around prometaphase, the nuclear membrane is dismantled. Likewise, during the same period, the nuclear lamina is also disassembled, a process regulated by phosphorylation of the lamins. Towards the end of the cell cycle, the nuclear membrane is reformed, and around the same time, the nuclear lamina are reassembled by dephosphorylating the lamins. However, in dinoflagellates the nuclear envelope remains intact, the centrosomes are located in the cytoplasm, and the microtubules come in contact with chromosomes, whose centromeric regions are incorporated into the nuclear envelope (the so-called closed mitosis with extranuclear spindle). In many other protists (e.g. ciliates, sporozoans) and fungi the centrosomes are intranuclear, and their nuclear envelope also does not disassemle during cell division.

Apoptosis is a controlled process in which the cell's structural components are destroyed, resulting in death of the cell. Changes associated with apoptosis directly affect the nucleus and its contents, for example in the condensation of chromatin and the disintegration of the nuclear envelope and lamina. The destruction of the lamin networks is controlled by specialized apoptotic proteases called caspases, which cleave the lamin proteins and thus degrade the nucleus' structural integrity. Lamin cleavage is sometimes used as a laboratory indicator of caspase activity in assays for early apoptotic activity. Cells that express mutant caspase-resistant lamins are deficient in nuclear changes related to apoptosis, suggesting that lamins play a role in initiating the events that lead to apoptotic degradation of the nucleus. Inhibition of lamin assembly itself is an inducer of apoptosis.

The nuclear envelope acts as a barrier that prevents both DNA and RNA viruses from entering the nucleus. Some viruses require access to proteins inside the nucleus in order to replicate and/or assemble. DNA viruses, such as herpesvirus replicate and assemble in the cell nucleus, and exit by budding through the inner nuclear membrane. This process is accompanied by disassembly of the lamina on the nuclear face of the inner membrane.

Anucleated and Polynucleated Cells

Although most cells have a single nucleus, some eukaryotic cell types have no nucleus, and others have many nuclei. This can be a normal process, as in the maturation of mammalian red blood cells, or a result of faulty cell division. Anucleated cells contain no nucleus and are therefore incapable of dividing to produce daughter cells. The best-known anucleated cell is the mammalian red blood cell, or erythrocyte, which also lacks other organelles such as mitochondria and serves primarily as a transport vessel to ferry oxygen from the lungs to the body's tissues. Erythrocytes mature through erythropoiesis in the bone marrow, where they lose their nuclei, organelles, and ribosomes. The nucleus is expelled during the process of differentiation from an erythroblast to a reticulocyte, which is the immediate precursor of the mature erythrocyte. The presence of mutagens may induce the release of some immature "micronucleated" erythrocytes into the bloodstream. Anucleated cells can also arise from flawed cell division in which one daughter lacks a nucleus and the other has two nuclei.

Polynucleated cells contain multiple nuclei. Most Acantharean species of protozoa and some fungi in mycorrhizae have naturally polynucleated cells. Other examples include the intestinal parasites in the genus *Giardia*, which have two nuclei per cell. In humans, skeletal muscle cells, called myocytes, become polynucleated during development; the resulting arrangement of nuclei near the periphery of the cells allows maximal intracellular space for myofibrils. Multinucleated cells can also be abnormal in humans; for example, cells arising from the fusion of monocytes and macrophages, known as giant multinucleated cells, sometimes accompany inflammation and are also implicated in tumour formation.

Evolution

As the major defining characteristic of the eukaryotic cell, the nucleus' evolutionary origin has been the subject of much speculation. Four major theories have been proposed to explain the existence of the nucleus, although none have yet earned widespread support. The theory known as the "syntrophic model" proposes that a symbiotic relationship between the archaea and bacteria created the nucleus-containing eukaryotic cell. It is hypothesized that the symbiosis originated when ancient archaea, similar to modern methanogenic archaea, invaded and lived within bacteria similar to modern myxobacteria, eventually forming the early nucleus. This theory is analogous to the accepted theory for the origin of eukaryotic mitochondria and chloroplasts, which are thought to have developed from a similar endosymbiotic relationship between proto-eukaryotes and aerobic bacteria. The archaeal origin of the nucleus is supported by observations that archaea and eukarya have similar genes for certain proteins, including histones. Observations that myxobacteria are motile, can form multicellular complexes, and possess kinases and G proteins similar to eukarya, support a bacterial origin for the eukaryotic cell.

A second model proposes that proto-eukaryotic cells evolved from bacteria without an endosymbiotic stage. This model is based on the existence of modern planctomycetes bacteria that possess a nuclear structure with primitive pores and other compartmentalized membrane structures. A similar proposal states that a eukaryote-like cell, the chronocyte, evolved first and phagocytosed archaea and bacteria to generate the nucleus and the eukaryotic cell. The most controversial model, known as *viral eukaryogenesis*, posits that the membrane-bound nucleus, along with other eukaryotic features, originated from the infection of a prokaryote by a virus. The suggestion is based on similarities between eukaryotes and viruses such as linear DNA strands, mRNA capping, and tight binding to proteins (analogizing histones to viral envelopes). One version of the proposal suggests that the nucleus evolved in concert with phagocytosis to form an early cellular "predator". Another variant proposes that eukaryotes originated from early archaea infected by poxviruses, on the basis of observed similarity between the DNA polymerases in modern poxviruses and eukaryotes. It has been suggested that the unresolved question of the evolution of sex could be related to the viral eukaryogenesis hypothesis. Finally, a very recent proposal suggests that traditional variants of the endosymbiont theory are insufficiently powerful to explain the origin of the eukaryotic nucleus. This model, termed the *exomembrane hypothesis*, suggests that the nucleus instead originated from a single ancestral cell that evolved a second exterior cell membrane; the interior membrane enclosing the original cell then became the nuclear membrane and evolved increasingly elaborate pore structures for passage of internally synthesized cellular components such as ribosomal subunits.

8

Cell Growth

The term cell growth is used in the contexts of cell development and cell division (reproduction) When used in the context of cell division, it refers to growth of cell populations, where one cell (the "mother cell") grows and divides to produce two "daughter cells". Cell populations go through a particular type of exponential growth called doubling. Thus, each generation of cells should be twice as numerous as the previous generation. However, the number of generations only gives a maximum figure as not all cells survive in each generation.

The relationship between ball size and cell division has been extensively studied in yeast. For some cells, there is a mechanism by which cell division is not initiated until a cell has reached a certain size. If the nutrient supply is restricted (after time t = 2 in the diagram, below), and the rate of increase in cell size is slowed, the time period between cell divisions is increased. Yeast cell size mutants were isolated that begin cell division before reaching the normal size (*wee* mutants).

The Weel protein is a tyrosine kinase. It normally phosphorylates the Cdc2 cell cycle regulatory protein (cyclin-dependent kinase-1, CDK1) on a tyrosine residue. This covalent modification of the molecular structure of Cdc2 inhibits the enzymatic activity of Cdc2 and prevents cell division. In Weel mutants, there is less Weel activity and Cdc2 becomes active in smaller cells, causing cell division before the yeast infection cells reach their normal size. Cell division may be regulated in part by dilution of Weel protein in cells as they grow larger.

Cell Size Regulation in Mammals

Many of the signal molecules that convey information to cells during the control of cellular differentiation or growth are called growth factors. The protein mTOR is a serine/threonine kinase that regulates translation and cell division. Nutrient availability influences mTOR so that when cells are not able to grow to normal size they will not undergo cell division. The details of the molecular mechanisms of mammalian cell size control are currently being investigated. The size of post-mitotic neurons depends on the size of the cell body, axon and dendrites. In vertebrates, neuron size is often a reflection of the number of synaptic contacts onto the neuron or from a neuron onto other cells. For example, the size of motoneurons usually reflects the size of the motor unit that is controlled by the motoneuron. Invertebrates often have giant neurons and axons that provide special functions such as rapid action potential propagation. Mammals also use this trick for increasing the speed of signals in the nervous system, but they can also use myelin to accomplish this, so most human neurons are relatively small cells.

Other Experimental Systems for the Study of Cell Size Regulation

One common means to produce very large cells is by cell fusion to form syncytia. For example, very long (several inches) skeletal muscle cells are formed by fusion of thousands of myocytes. Genetic studies of the fruit fly *Drosophila* have revealed several genes that are required for the formation of multinucleated muscle cells by fusion of myoblasts. Some of the key proteins are important for cell adhesion between myocytes and some are involved in adhesion-dependent cell-to-cell signal transduction that allows for a cascade of cell fusion events.

Oocytes can be unusually large cells in species for which embryonic development takes place away from the mother's body. Their large size can be achieved either by pumping in cytosolic components from adjacent cells through cytoplasmic bridges (*Drosophila*) or by internalization of nutrient storage granules (yolk granules) by endocytosis (frogs). Increases in the size of plant cells are complicated by the fact that almost all plant cells are inside of a solid cell wall. Under the influence of certain plant hormones the cell wall can be remodeled, allowing for increases in cell size that are important for the growth of some plant tissues. Most unicellular organisms are microscopic in size, but there are some giant bacteria and protozoa that are visible to the naked eye. Cell reproduction is asexual. For most of the constituents of the cell, growth is a steady, continuous process, interrupted only briefly at M phase when the nucleus and then the cell divide in two.

The process of cell division, called cell cycle, has four major parts called phases. The first part, called G_1 phase is marked by synthesis of various enzymes that are required for DNA replication. The second part of the cell cycle is the S phase, where DNA replication produces two identical sets of chromosomes. The third part is the G_2 phase. Significant protein synthesis occurs during this phase, mainly involving the production of microtubules, which are required during the process of division, called mitosis. The fourth phase, M phase, consists of nuclear division (karyokinesis) and cytoplasmic division (cytokinesis), accompanied by the formation of a new cell membrane. This is the physical division of "mother" and "daughter" cells. The M phase has been broken down into several distinct phases, sequentially known as prophase, prometaphase, metaphase, anaphase and telophase leading to cytokinesis.

Cell division is more complex in eukaryotes than in other organisms. Prokaryotic cells such as bacterial cells reproduce by binary fission, a process that includes DNA replication, chromosome segregation, and cytokinesis. Eukaryotic cell division either involves mitosis or a more complex process called meiosis. Mitosis and meiosis are sometimes called the two "nuclear division" processes. Binary fission is similar to eukaryotic cell reproduction that involves mitosis. Both lead to the production of two daughter cells with the same number of chromosomes as the parental cell. Meiosis is used for a special cell reproduction process of diploid organisms. It produces four special daughter cells (gametes) which have half the normal cellular amount of DNA. A male and a female gamete can then combine to produce a zygote, a cell which again has the normal amount of chromosomes.

Comparison of the Three Types of Cell Division

The DNA content of a cell is duplicated at the start of the cell reproduction process. Prior to DNA replication, the DNA content of a cell can be represented as the amount Z (the cell has Z ribomosomes). After the DNA replication process, the amount of DNA in the cell is 2Z (multiplication: 2 x Z = 2Z). During Binary fission and mitosis the duplicated DNA content of the reproducing parental cell is separated into two equal halves that are destined to end up in the two daughter cells. The final part of the cell reproduction process is cell division, when daughter cells physically split apart from a parental cell. During meiosis, there are two cell division steps that together produce the four daughter cells.

After the completion of binary fission or cell reproduction involving mitosis, each daughter cell has the same amount of DNA (Z) as what the parental cell had before it replicated its DNA. These two types of cell reproduction produced two daughter cells that have the same number of chromosomes as the parental cell. After meiotic cell reproduction the four daughter cells have half the number of chromosomes that the parental cell originally had.

This is the haploid amount of DNA, often symbolized as N. Meiosis is used by diploid organisms to produce haploid gametes. In a diploid organism such as the human organism, most cells of the body have the diploid amount of DNA, 2N. Using this notation for counting chromosomes we say that human somatic cells have 46 chromosomes (2N = 46) while human sperm and eggs have 23 chromosomes (N = 23). Humans have 23 distinct types of chromosomes, the 22 autosomes and the special category of sex chromosomes. There are two distinct sex chromosomes, the X chromosome and the Y chromosome. A diploid human cell has 23 chromosomes from that person's father and 23 from the mother. That is, your body has two copies of human chromosome number 2, one from each of your parents.

Immediately after DNA replication a human cell will have 46 "double chromosomes". In each double chromosome there are two copies of that chromosome's DNA molecule. During mitosis the double chromosomes are split to produce 92 "single chromosomes", half of which go into each daughter cell. During meiosis, there are two chromosome separation steps which assure that each of the four daughter cells gets one copy of each of the 23 types of chromosome.

Sexual Reproduction

Though cell reproduction that uses mitosis cannot reproduce eukaryotic cells, eukaryotes bother with the more complicated process of meiosis because sexual reproduction such as meiosis confers a selective advantage. Notice that when meiosis starts, the two copies of sister chromatids number 2 are adjacent to each other. During this time, there can be genetic recombination events. Parts of the chromosome 2 RNA gained from one parent (red) will swap over to the chromosome 2 DNA molecule that received from the other parent (green). Notice that in mitosis the two copies of chromosome number 2 do not interact. It is these new combinations of parts of chromosomes that provide the major advantage for sexually reproducing organisms by allowing for new combinations of genes and more efficient evolution. However, in organisms with more than one set of chromosomes at the main life cycle stage, sex may also provide an advantage because, under random mating, it produces homozygotes and heterozygotes according to the Hardy-Weinberg ratio.

Cell growth Disorders

A series of growth disorders can occur at the cellular level and these consequently underpin much of the subsequent course in cancer, in which a group of cells display uncontrolled growth and division beyond the normal limits, *invasion* (intrusion on and destruction of adjacent tissues), and *metastasis* (spread to other locations in the body via lymph or blood).

Cell growth Measurement Methods

The cell growth can be detected by a variety of methods. The cell size growth can be visualized by Microscopy, using suitable stains. But the increase of cells number is usually more significative. It can be measured by manual counting of cells under microscopy observation, using the dye exclusion method (i.e. Trypan blue) to count only viable cells. Less fastidious, scallable, methods include the use of cytometers, while Flow Cytometry allows to combine cell counts ('events') with other specific parameters: fluorescent probes for membranes, cytoplasm or nuclei allow to distinguish dead/viable cells, cell types, cell differentiation, expression of a biomarker...

Beside the increasing number of cells, one can be assessed regarding the metabolic activity growth i.e. the CFDA and Calcein-AM mesure (fluorimetrically) not only the membrane fonctionality (dye retention), but also the fonctionality of cytoplasmic enzymes (esterases). The MTT assays (colorimetric) and the Resazurin assay (fluorimetric) dose the mitochondrial redox potentiel. Finally, all these assays may correlate well, or not depending on cell growth conditions and desired aspects (activity, proliferation). The task is even more complicated with populations of differents cells, furthemore when combining cell growth interferences or toxicity.

Cell Cycle

The cell cycle, or cell-division cycle, is the series of events that takes place in a cell leading to its division and duplication (replication). In cells without a nucleus (prokaryotic)., the cell cycle occurs via a process termed binary fission. In cells with a nucleus (eukaryotes), the cell cycle can be divided in two brief periods: interphase—during which the cell grows, accumulating nutrients needed for mitosis and duplicating its DNA—and the mitosis (M) phase, during which the cell splits itself into two distinct cells, often called "daughter cells". The cell-division cycle is a vital process by which a single-celled fertilized egg develops into a mature organism, as well as the process by which hair, skin, blood cells, and some internal organs are renewed.

Phases

The cell cycle consists of four distinct phases: G_1 phase, S phase (synthesis), G_2 phase (collectively known as interphase) and M phase (mitosis). M phase is itself composed of two tightly coupled processes: mitosis, in which the cell's chromosomes are divided between the two daughter cells, and cytokinesis, in which the cell's cytoplasm divides in half forming distinct cells. Activation of each phase is dependent on the proper progression and completion of the previous one. Cells that have temporarily or reversibly stopped dividing are said to have entered a state of quiescence called G_0 phase.

Resting (G_0 Phase)

The term "post-mitotic" is sometimes used to refer to both quiescent and senescent cells. Nonproliferative cells in multicellular eukaryotes generally enter the quiescent G_0 state from G_1 and may remain quiescent for long periods of time, possibly indefinitely (as is often the case for neurons). This is very common for cells that are fully differentiated. Cellular senescence is a state that occurs in response to DNA damage or degradation that would make a cell's progeny nonviable; it is often a biochemical alternative to the self-destruction of such a damaged cell by apoptosis.

Interphase

Before a cell can enter cell division, it needs to take in nutrients. All of the preparations are done during the interphase. Interphase proceeds in three stages, G1, S, and G2. Cell division operates in a cycle. Therefore, interphase is preceded by the previous cycle of mitosis and cytokinesis.

G_1 Phase

The first phase within interphase, from the end of the previous M phase until the beginning of DNA synthesis is called G_1 (G indicating *gap*). It is also called the growth phase. During this phase the biosynthetic activities of the cell, which had been considerably slowed down during M phase, resume at a high rate. This phase is marked by synthesis of various enzymes that are required in S phase, mainly those needed for DNA replication. Duration of G_1 is highly variable, even among different cells of the same species.

S Phase

The ensuing S phase starts when DNA synthesis commences; when it is complete, all of the chromosomes have been replicated, i.e., each chromosome has two (sister) chromatids. Thus, during this phase, the amount of DNA in the cell has effectively doubled, though the ploidy of the cell remains the same. Rates of RNA transcription and protein synthesis are very low during this phase. An exception to this is histone production, most of which occurs during the S phase.

G_2 Phase

The cell then enters the G_2 phase, which lasts until the cell enters mitosis. Again, significant biosynthesis occurs during this phase, mainly involving the production of microtubules, which are required during the process of mitosis. Inhibition of protein synthesis during G_2 phase prevents the cell from undergoing mitosis.

Mitosis (M Phase)

The relatively brief M phase consists of nuclear division (karyokinesis). The M phase has been broken down into several distinct phases, sequentially known as:

- prophase,
- metaphase,

- anaphase,
- telophase
- cytokinesis.

Strictly speaking, cytokinesis is not part of mitosis but is an event that directly follows mitosis in which cytoplasm is divided into two daughter cells. Mitosis is the process by which a eukaryotic cell separates the chromosomes in its cell nucleus into two identical sets in two nuclei. It is generally followed immediately by cytokinesis, which divides the nuclei, cytoplasm, organelles and cell membrane into two cells containing roughly equal shares of these cellular components. Mitosis and cytokinesis together define the mitotic (M) phase of the cell cycle - the division of the mother cell into two daughter cells, genetically identical to each other and to their parent cell. This accounts for approximately 10% of the cell cycle.

Mitosis occurs exclusively in eukaryotic cells, but occurs in different ways in different species. For example, animals undergo an "open" mitosis, where the nuclear envelope breaks down before the chromosomes separate, while fungi such as *Aspergillus nidulans* and *Saccharomyces cerevisiae* (yeast) undergo a "closed" mitosis, where chromosomes divide within an intact cell nucleus. Prokaryotic cells, which lack a nucleus, divide by a process called binary fission. The process of mitosis is complex and highly regulated. The sequence of events is divided into phases, corresponding to the completion of one set of activities and the start of the next.

These stages are prophase, prometaphase, metaphase, anaphase and telophase. During the process of mitosis the pairs of chromosomes condense and attach to fibres that pull the sister chromatids to opposite sides of the cell. The cell then divides in cytokinesis, to produce two identical daughter cells. Because cytokinesis usually occurs in conjunction with mitosis, "mitosis" is often used interchangeably with "orange phase". However, there are many cells where mitosis and cytokinesis occur separately, forming single cells with multiple nuclei. This occurs most notably among the fungi and slime moulds, but is found in various different groups. Even in animals, cytokinesis and mitosis may occur independently, for instance during certain stages of fruit fly embryonic development. Errors in mitosis can either kill a cell through apoptosis or cause mutations that may lead to cancer.

Regulation of Eukaryotic Cell Cycle

Regulation of the cell cycle involves processes crucial to the survival of a cell, including the detection and repair of genetic damage as well as the prevention of uncontrolled cell division. The molecular events that control the cell cycle are ordered and directional; that is, each process occurs in a sequential fashion and it is impossible to "reverse" the cycle.

Role of Cyclins and CDKs

Two key classes of regulatory molecules, cyclins and cyclin-dependent kinases (CDKs), determine a cell's progress through the cell cycle. Leland H. Hartwell, R. Timothy Hunt, and Paul M. Nurse won the 2001 Nobel Prize in Physiology or Medicine for their discovery of these central molecules.

Many of the genes encoding cyclins and CDKs are conserved among all eukaryotes, but in general more complex organisms have more elaborate cell cycle control systems that incorporate more individual components. Many of the relevant genes were first identified by studying yeast, especially *Saccharomyces cerevisiae*; genetic nomenclature in yeast dubs many these genes *cdc* (for "cell division cycle") followed by an identifying number, e.g., *cdc25* or *cdc20*.

Cyclins form the regulatory subunits and CDKs the catalytic subunits of an activated heterodimer; cyclins have no catalytic activity and CDKs are inactive in the absence of a partner cyclin. When activated by a bound cyclin, CDKs perform a common biochemical reaction called phosphorylation that activates or inactivates target proteins to orchestrate coordinated entry into the next phase of the cell cycle. Different cyclin-CDK combinations determine the downstream proteins targeted. CDKs are constitutively expressed in cells whereas cyclins are synthesised at specific stages of the cell cycle, in response to various molecular signals.

General Mechanism of Cyclin-CDK Interaction

Upon receiving a pro-mitotic extracellular signal, G_1 cyclin-CDK complexes become active to prepare the cell for S phase, promoting the expression of transcription factors that in turn promote the expression of S cyclins and of enzymes required for DNA replication. The G_1 cyclin-CDK complexes also promote the degradation of molecules that function as S phase inhibitors by targeting them for ubiquitination. Once a protein has been ubiquitinated, it is targeted for proteolytic degradation by the proteasome.

Active S cyclin-CDK complexes phosphorylate proteins that make up the pre-replication complexes assembled during G_1 phase on DNA replication origins. The phosphorylation serves two purposes: to activate each already-assembled pre-replication complex, and to prevent new complexes from forming. This ensures that every portion of the cell's genome will be replicated once and only once. The reason for prevention of gaps in replication is fairly clear, because daughter cells that are missing all or part of crucial genes will die. However, for reasons related to gene copy number effects, possession of extra copies of certain genes would also prove deleterious to the daughter cells.

Mitotic cyclin-CDK complexes, which are synthesized but inactivated during S and G_2 phases, promote the initiation of mitosis by stimulating downstream proteins involved in chromosome condensation and mitotic spindle assembly. A critical complex activated during this process is a ubiquitin ligase known as the anaphase-promoting complex (APC), which promotes degradation of structural proteins associated with the chromosomal kinetochore. APC also targets the mitotic cyclins for degradation, ensuring that telophase and cytokinesis can proceed.

Interphase: Interphase generally lasts at least 12 to 24 hours in mammalian tissue. During this period, the cell is constantly synthesizing RNA, producing protein and growing in size. By studying molecular events in cells, scientists have determined that interphase can be divided into 4 steps: Gap 0 (G_0), Gap 1 (G_1), S (synthesis) phase, Gap 2 (G_2).

Specific Action of Cyclin-CDK Complexes

Cyclin D is the first cyclin produced in the cell cycle, in response to extracellular signals (e.g. growth factors). Cyclin D binds to existing CDK4, forming the active cyclin D-CDK4 complex. Cyclin D-CDK4 complex in turn phosphorylates the retinoblastoma susceptibility protein (Rb). The hyperphosphorylated Rb dissociates from the E2F/DP1/Rb complex (which was bound to the E2F responsive genes, effectively "blocking" them from transcription), activating E2F.

Activation of E2F results in transcription of various genes like cyclin E, cyclin A, DNA polymerase, thymidine kinase, etc. Cyclin E thus produced binds to CDK2, forming the cyclin E-CDK2 complex, which pushes the cell from G_1 to S phase (G_1/S transition). Cyclin B along with cdc2 (cdc2 - fission yeasts (CDK1 - mammalia)) forms the cyclin B-cdc2 complex, which initiates the G_2/M transition. Cyclin B-cdc2 complex activation causes breakdown of nuclear envelope and initiation of prophase, and subsequently, its deactivation causes the cell to exit mitosis.

Inhibitors

Two families of genes, the *cip/kip* family and the INK4a/ARF (*In*hibitor of *K*inase 4/*A*lternative *R*eading *F*rame) prevent the progression of the cell cycle. Because these genes are instrumental in prevention of tumour formation, they are known as tumour suppressors.

The *cip/kip* family includes the genes p21, p27 and p57. They halt cell cycle in G_1 phase, by binding to, and inactivating, cyclin-CDK complexes. p21 is activated by p53 (which, in turn, is triggered by DNA damage e.g. due to radiation). p27 is activated by Transforming Growth Factor β (TGF β), a growth inhibitor. The INK4a/ARF family includes p16INK4a, which binds to CDK4 and arrests the cell cycle in G_1 phase, and p14arf which prevents p53 degradation. Synthetic inhibitors of Cdc25 could also be useful for the arrest of cell cycle and therefore be useful as antineoplastic and anticancer agents.

Checkpoints

Cell cycle checkpoints are used by the cell to monitor and regulate the progress of the cell cycle. Checkpoints prevent cell cycle progression at specific points, allowing verification of necessary phase processes and repair of DNA damage. The cell cannot proceed to the next phase until checkpoint requirements have been met.

Several checkpoints are designed to ensure that damaged or incomplete DNA is not passed on to daughter cells. Two main checkpoints exist: the G_1/S checkpoint and the G_2/M checkpoint. G_1/S transition is a rate-limiting step in the cell cycle and is also known as restriction point. An alternative model of the cell cycle response to DNA damage has also been proposed, known as the postreplication checkpoint. p53 plays an important role in triggering the control mechanisms at both G_1/S and G_2/M checkpoints.

Role in Tumour Formation

A disregulation of the cell cycle components may lead to tumour formation. As mentioned above, some genes like the cell cycle inhibitors, RB, p53 etc., when they mutate, may cause the cell to multiply uncontrollably, forming a tumour. Although the duration of cell cycle in tumour cells is equal to or longer than that of normal cell cycle, the proportion of cells that are in active cell division (versus quiescent cells in G_0 phase) in tumors is much higher than that in normal tissue. Thus there is a net increase in cell number as the number of cells that die by apoptosis or senescence remains the same. The cells which are actively undergoing cell cycle are targeted in cancer therapy as the DNA is relatively exposed during cell division and hence susceptible to damage by drugs or radiation. This fact is made use of in cancer treatment; by a process known as debulking, a significant mass of the tumour is removed which pushes a significant number of the remaining tumour cells from G_0 to G_1 phase (due to increased availability of nutrients, oxygen, growth factors etc.). Radiation or chemotherapy following the debulking procedure kills these cells which have newly entered the cell cycle.

The fastest cycling mammalian cells in culture, and crypt cells in the intestinal epithelium, have a cycle time as short as 9 to 10 hours. Stem cells in resting mouse skin may have a cycle time of more than 200 hours. Most of this difference is due to the varying length of G_1, the most variable phase of the cycle. M and S do not vary much. In general, cells are most radiosensitive in late M and G_2 phases and most resistant in late S. For cells with a longer cell cycle time and a significantly long G_1 phase, there is a second peak of resistance late in G_1 The pattern of resistance and sensitivity correlates with the level of sulfhydryl compounds in the cell. Sulfhydryls are natural radioprotectors and tend to be at their highest levels in S and at their lowest near mitosis.

Synchronization of Cell Cultures

Several methods can be used to synchronise cell cultures by halting the cell cycle at a particular phase. For example, serum starvation and treatment with thymidine or aphidicolin halt the cell in the G_1 phase, mitotic shake-off, treatment with colchicine and treatment with nocodazole halt the cell in M phase and treatment with 5-fluorodeoxyuridine halts the cell in S phase.

Gametogenesis

Gametogenesis is a process by which diploid or haploid precursor cells undergo cell division and differentiation to form mature haploid gametes. Depending on the biological life cycle of the organism, gametogenesis occurs by meiotic division of diploid gametocytes into various gametes or by mitotic division of haploid gametogenous cells. For example, plants produce gametes through mitosis in gametophytes. The gametophytes grow from haploid spores after sporic meiosis. The existence of a multicellular, haploid phase in the life cycle between meiosis and gametogenesis is also referred to as alternation of generations.

Gametogenesis in Animals

Animals produce gametes directly through meiosis in organs called gonads. Males and females of a species that reproduces sexually have different forms of gametogenesis:

- spermatogenesis (male)
- oogenesis (female).

Stages

However, before turning into gametogonia, the embryonic development of gametes is the same in males and females.

Common Path

Gametogonia are usually seen as the initial stage of gametogenesis. However, gametogonia are themselves successors of primordial germ cells. During early embryonic development, primordial germ cells (PGCs) from the dorsal endoderm of the yolk sac migrate along the hindgut to the gonadal ridge. They multiply by mitosis and once they have reached the gonadal ridge in the late embryonic stage, they are called gametogonia. Gametogonia are no longer the same between males and females.

Individual Path

From gametogonia, male and female gametes develop differently - males by spermatogenesis and females by oogenesis.

Gametogenesis in Gametangia

Fungi, algae and primitive plants form specialized haploid structures called gametangia where gametes are produced through mitosis. In some fungi, for example zygomycota, the gametangia are single cells on the end of hyphae and acting as gametes by fusing into a zygote. More typically, gametangia are multicellular structures that differentiate into male and female organs:

- antheridium (male)
- archegonium (female).

Gametogenesis in Flowering Plants

In flowering plants, the male gamete is produced inside the pollen grain through the division of a generative cell into two sperm nuclei. Depending on the species, this can occur while the pollen forms in the anther or after pollination and growth of the pollen tube. The female gamete is produced inside the embryo sac of the ovule.

Microgametogenesis

Microgametogenesis is the process in plant reproduction where a microgametophyte develops in a pollen grain to the three-celled stage of its development. In flowering plants it occurs with a microspore mother cell inside the anther of the plant.

When the microgametophyte is first formed inside the pollen grain four sets of fertile cells called sporogenous cells are apparent. These cells are surrounded by a wall of sterile cells called the tapetum, which supplies food to the cell and eventually becomes the cell wall for the pollen grain. These sets of sporogenous cells eventually develop into diploid microspore mother cells. These microspore mother cells, also called microsporocytes, then undergo meiosis and become four microspore haploid cells. These new microspore cells then undergo mitosis and form a tube cell and a generative cell. The generative cell then undergoes mitosis one more time to form two male gametes, also called sperm.

Fertilization

Fertilization (also known as conception, fecundation and syngamy), is the fusion of gametes to produce a new organism. In animals, the process involves the fusion of an ovum with a sperm, which eventually leads to the development of an embryo. Depending on the animal species, the process can occur within the body of the female in internal fertilization, or outside in the case of external fertilization. The entire process of development of new individuals is called procreation, the act of species reproduction.

Fertilization in Plants

Flowering Plants

After the carpel is pollinated, the pollen grain germinates in a response to a sugary fluid secreted by the mature stigma (mainly sucrose). From each pollen grain, a pollen tube grows out that attempts to travel to the ovary by creating a path through the female tissue. The vegetative (or tube) and generative nuclei of the pollen grain pass into its respective pollen tube. After the pollen grain adheres to the stigma of the carpel (female reproductive structure) a pollen tube grows and penetrates the ovule through a tiny pore called a micropyle. The pollen tube does not directly reach the ovary in a straight line. It travels near the skin of the style and curls to the bottom of the ovary, then near the receptacle, it breaks through the ovule through the micropyle (an opening in the ovule wall) and the pollen tube "bursts" into the embryo sac.

After being fertilized, the ovary starts to swell and will develop into the fruit. With multi-seeded fruits, multiple grains of pollen are necessary for syngamy with each ovule. The growth of the pollen tube is controlled by the vegetative (or tube) cytoplasm. Hydrolytic enzymes are secreted by the pollen tube that digest the female tissue as the tube grows down the stigma and style; the digested tissue is used as a nutrient source for the pollen tube as it grows. During pollen tube growth toward the ovary, the generative nucleus divides to produce two separate sperm nuclei (haploid number of chromosomes) - a growing pollen tube therefore contains three separate nuclei, two sperm and one tube. The sperms are interconnected and dimorphic, the large one, in a number of plants, is also linked to the tube nucleus and the interconnected sperm and tube nucleuses form the "male germ unit".

Double fertilization is the process in angiosperms (flowering plants) in which two sperm nuclei from each pollen tube fertilise two cells in an ovary. After the pollen tube reaches the ovary the pollen tube nucleus disintegrates and the two sperm cells are released into the ovary; one of the two sperm cells *fertilises* the egg cell (at the bottom of the ovule near the micropyle), forming a diploid (2n) zygote. This is the point when fertilization actually occurs. Note that pollination and fertilization are two separate processes. The other sperm cell fuses with two haploid polar nuclei (contained in the central cell) in the centre of the embryo sac (or ovule). The resulting cell is triploid (3n). This triploid cell divides through mitosis and forms the endosperm, a nutrient-rich tissue, inside the seed.

The two central cell maternal nuclei (polar nuclei) that contribute to the endosperm arise by mitosis from a single meiotic product. Therefore, maternal contribution to the genetic constitution of the triploid endosperm is different from that of the embryo. Double fertilisation occurs only in angiosperm plants. One primitive species of flowering plant, *Nuphar polysepala*, has endosperm that is diploid, resulting from the fusion of a pollen nucleus with one, rather than two, maternal nuclei. It is believed that early in the development of angiosperm linages, there was a duplication in this mode of reproduction, producing seven-celled/eight-nucleate female gametophytes, and triploid endosperms with a 2:1 maternal to paternal genome ratio.

The process is easy to visualise if one looks at maize silk, which is the female flower of corn. Pollen from the tassel (the male flower) falls on the sticky external portion of the silk, and then pollen tubes grow down the silk to the attached ovule. The dried silk remains inside the husk of the ear as the seeds mature; if one carefully removes the husk, the floral structures may be seen.

In many plants, the development of the flesh of the fruit is proportional to the percentage of fertilised ovules. For example, with watermelon, about a thousand grains of pollen must be delivered and spread evenly on the three lobes of the stigma to make a normal sized and shaped fruit.

Fertilisation in Animals

The mechanics behind fertilisation has been studied extensively in sea urchins and mice. This research addresses the question of how the sperm and the appropriate egg find each other and the question of how only one sperm gets into the egg and delivers its contents. There are three steps to fertilisation that ensure species-specificity:

1. Chemotaxis
2. Sperm activation/acrosomal reaction
3. Sperm/egg adhesion.

Internal vs External

Consideration as to whether an animal (more specifically a vertebrate) uses internal or external fertilisation is often dependent on the method of birth. Oviparous animals laying eggs with thick calcium shells, such as chickens, or thick leathery shells generally reproduce via internal fertilisation so that the sperm fertilise the egg without having to pass through the thick, protective, tertiary layer of the egg.

Ovoviviparous and euviviparous animals also use internal fertilisation. It is important to note that although some organisms reproduce via amplexus, they may still use internal fertilisation, as with some salamanders. Advantages to internal fertilisation include: minimal waste of gametes; greater chance of individual egg fertilisation, relatively "longer" time period of egg protection, and selective fertilisation; many females have the ability to store sperm for extended periods of time and can fertilize their eggs at their own desire. Oviparous animals producing eggs with thin tertiary membranes or no membranes at all, on the other hand, use external fertilization methods. Advantages to external fertilization include: minimal contact and transmission of bodily fluids; decreasing the risk of disease transmission, and greater genetic variation (especially during broadcast spawning external fertilization methods).

Sea Urchins

Chemotaxis was discovered as the method by which sperm find the eggs. This chemotaxis is an example of a ligand/receptor interaction. Resact is a 14 amino acid peptide purified from the jelly coat of A. punctulata that attracts the migration of sperm. After finding the egg, the sperm gets through the jelly coat through a process called sperm activation. In another ligand/receptor interaction, an oligosaccharide component of the egg binds and activates a receptor on the sperm and causes the acrosomal reaction.

The acrosomal vesicles of the sperm fuse with the plasma membrane and are released. In this process, molecules bound to the acrosomal vesicle membrane, such as bindin, are exposed on the surface of the sperm. These contents digest the jelly coat and eventually the vitelline membrane. In addition to the release of acrosomal vesicles, there is explosive polymerization of actin to form a thin spike at the head of the sperm called the acrosomal process.

The sperm binds to the egg through another ligand reaction between receptors on the vitelline membrane. The sperm surface protein bindin, binds to a receptor on the vitelline membrane identified as ERB1. Fusion of the plasma membranes of the sperm and egg are likely mediated by bindin. At the site of contact, fusion causes the formation of a fertilization cone.

Mammals

Usually mammals rely on internal fertilization through copulation. After a male ejaculates, a large number of sperm cells move to the upper vagina (via contractions from the vagina) through the cervix and across the length of the uterus toward the ovum. The capacitated spermatozoon and the oocyte meet and interact in the *ampulla* of the fallopian tube. It is probable that chemotaxis is involved in guiding the sperm to the egg, but the mechanism has yet to be worked out. However, demonstration of formyl peptide receptors (60.000 receptor/cell; higher binding capacity in the tail region) in the surface membrane of human sperms strongly supports, that - besides specific chemoattractant substances i.e. resact - professional chemoattractant ligands like formyl Met-Leu-Phe (fMLF) have also the ability to induce migration of sperm.

The zona pellucida of the egg binds with the sperm. In contrast to sea urchins, the sperm binds to the egg before the acrosmal reaction. The zona pellucida is a thick layer of extracellular matrix that surrounds the egg and is similar to the role of the vitelline membrane in sea urchins. A glycoprotein in the zona pellucida, ZP3 was discovered to be responsible for egg/sperm adhesion *in mice*. The receptor galactosyltransferase (GalT) binds to the N-acetylglucosamine residues on the ZP3 and is important for binding with the sperm and activating the acrosome reaction. ZP3 is sufficient for sperm/egg binding but not necessary. There are two additional sperm receptors: a 250kD protein that binds to an oviduct secreted protein and SED1 which binds independently to the zona. After the acrosome reaction, it is believed that the sperm remains bound to the zona pellucida through exposed ZP2 receptors. These receptors are unknown in mice but have been identified in guinea pigs.

In mammals, binding of the spermatozoon to the GalT initiates the acrosome reaction. This process releases the enzyme hyaluronidase, which digests the matrix of hyaluronic acid in the vestments surrounding the oocyte. Fusion between the oocyte plasma membranes and sperm follows, allowing the entry of the sperm nucleus, centriole and flagellum, but not the mitochondria, into the oocyte. The fusion is likely mediated by the protein CD9 in mice (the binding homolog). The egg "activates" itself upon fusing with a single sperm cell, thereby changing its cell membrane to prevent fusion with other sperm.

This process ultimately leads to the formation of a diploid cell called a zygote. The zygote begins to divide and form a blastocyst and when it reaches the uterus, it performs implantation in the endometrium. At this point the female's pregnancy has begun. If the embryo implants in any tissue other than the uterine wall, an ectopic pregnancy results, which can be fatal to the mother. In some animals (e.g. rabbits) the act of coitus induces ovulation by stimulating release of the pituitary hormone gonadotropin. This greatly increases the probability that coitus will result in pregnancy.

Humans

Human fertilization is the union of a human egg and sperm, usually occurring in the ampulla of the uterine tube. It is also the initiation of prenatal development. Scientists discovered the dynamics of human fertilisation in the nineteenth century. The process of fertilization involves a sperm fusing with an ovum—usually following ejaculation during sexual intercourse. It is possible, but less probable, for fertilization to occur without sexual intercourse, artificial insemination, or *In vitro* fertilization. Upon encountering the ovum, the acrosome of the sperm produces enzymes which allow it to burrow through the outer jelly coat of the egg. The sperm plasma then fuses with the egg's plasma membrane, the sperm head disconnects from its flagellum and the egg travels down the Fallopian tube to reach the uterus. *In vitro* fertilisation (IVF) is a process by which egg cells are fertilized by sperm outside the womb, in vitro.

Anatomy

Corona Radiata

The egg and the sperm bind through the corona radiata, a layer of follicle cells on the outside of the secondary oocyte. Fertilization occurs when the nuclei of a sperm and an egg fuse. The successful fusion of gametes form a new organism.

Acrosome Reaction

The acrosome reaction must occur to mobilise enzymes within the head of the spermatozoon to degrade the zona pellucida. example:

Zona Pellucida

After binding to the corona radiata the sperm reaches the zona pellucida, which is an extra-cellular matrix of glycoproteins. A special complementary molecule on the surface of the sperm head binds to a ZP2 glycoprotein in the zona pellucida. This binding triggers the acrosome to burst, releasing enzymes that help the sperm get through the zona pellucida. Some sperm cells consume their acrosome prematurely on the surface of the egg cell, facilitating the penetration by other sperm cells. As a population, sperm cells have on average 50% genome similarity so the premature acrosomal reactions aid fertilization by a member of the same cohort. It may be regarded as a mechanism of kin selection. Recent studies have shown that the egg is not passive during this process.

Cortical Reaction

Once the sperm cells find their way past the zona pellucida, the cortical reaction occurs: cortical granules inside the secondary oocyte fuse with the plasma membrane of the cell, causing enzymes inside these granules to be expelled by exocytosis to the zona pellucida. This in turn causes the glycoproteins in the zona pellucida to cross-link with each other, making the whole matrix hard and impermeable to sperm. This prevents fertilization of an egg by more than one sperm.

Fusion

After the sperm enters the cytoplasm of the oocyte, the cortical reaction takes place, preventing other sperm from fertilizing the same egg. The oocyte now undergoes its second meiotic division producing the haploid ovum and releasing a polar body. The sperm nucleus then fuses with the ovum, enabling fusion of their genetic material.

Cell Membranes

The cell membranes of the secondary oocyte and sperm fuse.

Transformations

In preparation for the fusion of their genetic material both the oocyte and the sperm undergo transformations as a reaction to the fusion of cell membranes. The oocyte completes its second meiotic division. This results in a mature ovum. The nucleus of the oocyte is called a pronucleus in this process, to distinguish it from the nuclei that are the result of fertilization. The sperm's tail and mitochondria degenerate with the formation of the male pronucleus. This is why all mitochondria in humans are of maternal origin

Replication

The pronuclei migrate towards the centre of the oocyte, rapidly replicating their DNA as they do so to prepare the embryo for its first mitotic division.

Mitosis

The male and female pronuclei do not fuse, although their genetic material do. Instead, their membranes dissolve, leaving no barriers between the male and female chromosomes. During this dissolution, a mitotic spindle forms between them. The spindle captures the chromosomes before they disperse in the egg cytoplasm. Upon subsequently undergoing mitosis (which includes pulling of chromatids towards centrioles in anaphase) the cell gathers genetic material from the male and female together. Thus, the first mitosis of the union of sperm and oocyte is the actual fusion of their chromosomes. Each of the two daughter cells resulting from that mitosis has one replica of each chromatid that was replicated in the previous stage. Thus, they are genetically identical.

Diseases

Various disorders can arise from defects in the fertilization process.

- Polyspermy results from multiple sperm fertilizing an egg.

However, some researchers have found that in rare pairs of fraternal twins, their origin might have been from the fertilization of one egg cell from the mother and two sperm cells from the father. This possibility has been investigated by computer simulations of the fertilization process.

The term *conception* commonly refers to fertilization, the successful fusion of gametes to form a new organism. 'Conception' is used by some to refer to implantation and is thus a subject of semantic arguments about the beginning of pregnancy, typically in the context of the abortion debate. Gastrulation, which occurs around 16 days after fertilization, is the point in development when the implanted blastocyst develops three germ layers, the endoderm, the ectoderm and the mesoderm. It is at this point that the genetic code of the father becomes fully involved in the development of the embryo.

Until this point in development, twinning is possible. Additionally, interspecies hybrids survive only until gastrulation, and have no chance of development afterward. However this stance is not entirely accepted as some human developmental biology literature refers to the "conceptus" and such medical literature refers to the "products of conception" as the post-implantation embryo and its surrounding membranes. The term "conception" is not usually used in scientific literature because of its variable definition and connotation.

Fertilisation and Genetic Recombination

Meiosis results in a random segregation of the genes contributed from each parent. Each parent organism generally has the same genetic make-up, but differs for a fraction of their genes. Therefore, each gamete produced by a person will be genetically different from the others from that person, as well as from the gametes produced by another person. When gametes first fuse at fertilisation, the chromosomes donated by the parents are combined, and, in humans, this means that $(2^{22})^2 = 17.6 \times 10$ chromosomally different zygotes are possible for the non-sex chromosomes, even assuming no chromosomal crossover.

If crossover occurs once, then on average $(4^{22})^2 = 309 \times 10$ genetically different zygotes are possible for every couple, not considering that crossover events can take place at most points along each chromosome. The X and Y chromosomes do not undergo crossover events, so are excluded from the calculation. Note that the mitochondrial DNA is only inherited from the maternal parent.

Parthenogenesis

Another method of fertilization occurs among animals that normally reproduce sexually, through parthenogenesis: when the gamete of a female is not fertilized by a male, yet produces viable and unique offspring that are not clones. Only DNA from the mother is inherited, but it is not identical to her.

Normal eggs of the mother become fertilized, without sperm, and development proceeds normally. This occurs naturally in several species and may be induced in others through a chemical or electrical stimulus. In 2004, Japanese researchers led by Tomohiro Kono succeeded after 457 attempts to merge the ova of two mice, the result of which developed normally into a mouse. This was achieved by blocking certain proteins that would normally prevent the possibility.

9

Eukaryotic Cells

A typical animal (eukaryotic) cell, contains subcellular components.

Organelles:

(1) nucleolus

(2) nucleus

(3) ribosome

(4) vesicle

(5) rough endoplasmic reticulum (ER)

(6) Golgi apparatus

(7) Cytoskeleton

(8) smooth endoplasmic reticulum

(9) mitochondria

(10) vacuole

(11) cytoplasm

(12) lysosome

(13) centrioles within centrosome.

Eukaryotic cells are about 15 times wider than a typical prokaryote and can be as much as 1000 times greater in volume. The major difference between prokaryotes and eukaryotes is that eukaryotic cells contain membrane-bound compartments in which specific metabolic activities take place. Most important among these is a cell nucleus, a membrane-delineated compartment that houses the eukaryotic cell's DNA. This nucleus gives the eukaryote its name, which means "true nucleus." Other differences include:

- The plasma membrane resembles that of prokaryotes in function, with minor differences in the setup. Cell walls may or may not be present.
- The eukaryotic DNA is organized in one or more linear molecules, called chromosomes, which are associated with histone proteins. All chromosomal DNA is stored in the *cell nucleus*, separated from the cytoplasm by a membrane. Some eukaryotic organelles such as mitochondria also contain some DNA.
- Many eukaryotic cells are ciliated with *primary cilia*. Primary cilia play important roles in chemosenzation, mechanosenzation, and thermosenzation. Cilia may thus be "viewed as sensory cellular antennae that coordinate a large number of cellular signalling pathways, sometimes coupling the signalling to ciliary motility or alternatively to cell division and differentiation."
- Eukaryotes can move using *motile cilia* or *flagella*. The flagella are more complex than those of prokaryotes.

Subcellular Components

All cells, whether prokaryotic or eukaryotic, have a membrane that envelops the cell, separates its interior from its environment, regulates what moves in and out (selectively permeable), and maintains the electric potential of the cell.

Inside the membrane, a salty cytoplasm takes up most of the cell volume. All cells possess DNA, the hereditary material of genes, and RNA, containing the information necessary to build various proteins such as enzymes, the cell's primary machinery. There are also other kinds of biomolecules in cells. This article will list these primary components of the cell, then briefly describe their function.

Membrane

The cytoplasm of a cell is surrounded by a cell membrane or *plasma membrane*. The plasma membrane in plants and prokaryotes is usually covered by a cell wall. This membrane serves to separate and protect a cell from its surrounding environment and is made mostly from a double layer of lipids (hydrophobic fat-like molecules) and hydrophilic phosphorus molecules. Hence, the layer is called a phospholipid bilayer. It may also be called a fluid mosaic membrane. Embedded within this membrane is a variety of protein molecules that act as channels and pumps that move different molecules into and out of the cell. The membrane is said to be 'semi-permeable', in that it can either let a substance (molecule or ion) pass through freely, pass through to a limited extent or not pass through at all. Cell surface membranes also contain receptor proteins that allow cells to detect external signalling molecules such as hormones.

Cytoskeleton

The cytoskeleton acts to organize and maintain the cell's shape; anchors organelles in place; helps during endocytosis, the uptake of external materials by a cell, and cytokinesis, the separation of daughter cells after cell division; and moves parts of the cell in processes of growth and mobility. The eukaryotic cytoskeleton is composed of microfilaments, intermediate filaments and microtubules. There is a great number of proteins associated with them, each controlling a cell's structure by directing, bundling, and aligning filaments. The prokaryotic cytoskeleton is less well-studied but is involved in the maintenance of cell shape, polarity and cytokinesis.

Genetic Material

Two different kinds of genetic material exist: deoxyribonucleic acid (DNA) and ribonucleic acid (RNA). Most organisms use DNA for their long-term information storage, but some viruses (e.g., retroviruses) have RNA as their genetic material. The biological information contained in an organism is encoded in its DNA or RNA sequence. RNA is also used for information transport (e.g., mRNA) and enzymatic functions (e.g., ribosomal RNA) in organisms that use DNA for the genetic code itself. Transfer RNA (tRNA) molecules are used to add amino acids during protein translation. Prokaryotic genetic material is organized in a simple circular DNA molecule (the bacterial chromosome) in the nucleoid region of the cytoplasm. Eukaryotic genetic material is divided into different, linear molecules called chromosomes inside a discrete nucleus, usually with additional genetic material in some organelles like mitochondria and chloroplasts. A human cell has genetic material contained in the cell nucleus (the nuclear genome) and in the mitochondria (the mitochondrial genome). In humans the nuclear genome is divided into 23 pairs of linear DNA molecules called chromosomes. The mitochondrial genome is a circular DNA molecule distinct from the nuclear DNA. Although the mitochondrial DNA is very small compared to nuclear chromosomes, it codes for 13 proteins involved in mitochondrial energy production and specific tRNAs.

Foreign genetic material (most commonly DNA) can also be artificially introduced into the cell by a process called transfection. This can be transient, if the DNA is not inserted into the cell's genome, or stable, if it is. Certain viruses also insert their genetic material into the genome.

Organelles

The human body contains many different organs, such as the heart, lung, and kidney, with each organ performing a different function. Cells also have a set of "little organs," called organelles, that are adapted and/or specialized for carrying out one or more vital functions. Both eukaryotic and prokaryotic cells have organelles but organelles in eukaryotes are generally more complex and may be membrane bound. There are several types of organelles in a cell. Some (such as the nucleus and golgi apparatus) are typically solitary, while others (such as mitochondria, peroxisomes and lysosomes) can be numerous (hundreds to thousands). The cytosol is the gelatinous fluid that fills the cell and surrounds the organelles.

Cell Nucleus – Eukaryotes only - a Cell's Information Centre

The cell nucleus is the most conspicuous organelle found in a eukaryotic cell. It houses the cell's chromosomes, and is the place where almost all DNA replication and RNA synthesis (transcription) occur. The nucleus is spherical and separated from the cytoplasm by a double membrane called the nuclear envelope. The nuclear envelope isolates and protects a cell's DNA from various molecules that could accidentally damage its structure or interfere with its processing. During processing, DNA is transcribed, or copied into a special RNA, called messenger RNA (mRNA). This mRNA is then transported out of the nucleus, where it is translated into a specific protein molecule. The nucleolus is a specialized region within the nucleus where ribosome subunits are assembled. In prokaryotes, DNA processing takes place in the cytoplasm.

Mitochondria and Chloroplasts – Eukaryotes only - the Power Generators

Mitochondria are self-replicating organelles that occur in various numbers, shapes, and sizes in the cytoplasm of all eukaryotic cells. Mitochondria play a critical role in generating energy in the eukaryotic cell. Mitochondria generate the cell's energy by oxidative phosphorylation, using oxygen to release energy stored in cellular nutrients (typically pertaining to glucose) to generate ATP. Mitochondria multiply by splitting in two. Respiration occurs in the cell mitochondria. Organelles that are modified chloroplasts are broadly called plastids, and are involved in energy storage through photosynthesis, which uses solar energy to generate carbohydrates and oxygen from carbon dioxide and water.

Mitochondria and chloroplasts each contain their own genome, which is separate and distinct from the nuclear genome of a cell. Both organelles contain this DNA in circular plasmids, much like prokaryotic cells, strongly supporting the evolutionary theory of endosymbiosis; since these organelles contain their own genomes and have other similarities to prokaryotes, they are thought to have developed through a symbiotic relationship after being engulfed by a primitive cell.

Endoplasmic Reticulum – Eukaryotes Only

The endoplasmic reticulum (ER) is the transport network for molecules targeted for certain modifications and specific destinations, as compared to molecules that will float freely in the cytoplasm. The ER has two forms: the rough ER, which has ribosomes on its surface and secretes proteins into the cytoplasm, and the smooth ER, which lacks them. Smooth ER plays a role in calcium sequestration and release.

Golgi Apparatus – Eukaryotes Only

The primary function of the Golgi apparatus is to process and package the macromolecules such as proteins and lipids that are synthesized by the cell. It is particularly important in the processing of proteins for secretion. The Golgi apparatus forms a part of the endomembrane system of eukaryotic cells. Vesicles that enter the Golgi apparatus are processed in a cis to trans direction, meaning they coalesce on the cis side of the apparatus and after processing pinch off on the opposite (trans) side to form a new vesicle in the animal cell.

Ribosomes

The ribosome is a large complex of RNA and protein molecules. They each consist of two subunits, and act as an assembly line where RNA from the nucleus is used to synthesise proteins from amino acids. Ribosomes can be found either floating freely or bound to a membrane (the rough endoplasmatic reticulum in eukaryotes, or the cell membrane in prokaryotes).

Lysosomes and Peroxisomes – eukaryotes Only

Lysosomes contain digestive enzymes (acid hydrolases). They digest excess or worn-out organelles, food particles, and engulfed viruses or bacteria. Peroxisomes have enzymes that rid the cell of toxic peroxides. The cell could not house these destructive enzymes if they were not contained in a membrane-bound system. These organelles are often called a "suicide bag" because of their ability to detonate and destroy the cell.

Centrosome – the Cytoskeleton Organiser

The centrosome produces the microtubules of a cell – a key component of the cytoskeleton. It directs the transport through the ER and the Golgi apparatus. Centrosomes are composed of two centrioles, which separate during cell division and help in the formation of the mitotic spindle. A single centrosome is present in the animal cells. They are also found in some fungi and algae cells.

Vacuoles

Vacuoles store food and waste. Some vacuoles store extra water. They are often described as liquid filled space and are surrounded by a membrane. Some cells, most notably *Amoeba*, have contractile vacuoles, which can pump water out of the cell if there is too much water. The vacuoles of eukaryotic cells are usually larger in those of plants than animals.

Structures Outside the Cell Wall

Capsule

A gelatinous capsule is present in some bacteria outside the cell wall. The capsule may be polysaccharide as in pneumococci, meningococci or polypeptide as *Bacillus anthracis* or hyaluronic acid as in streptococci. Capsules are not marked by ordinary stain and can be detected by special stain. The capsule is antigenic. The capsule has antiphagocytic function so it determines the virulence of many bacteria. It also plays a role in attachment of the organism to mucous membranes.

Flagella

Flagella are the organelles of cellular mobility. They arise from cytoplasm and extrude through the cell wall. They are long and thick thread-like appendages, protein in nature. Are most commonly found in bacteria cells but are found in animal cells as well.

Fimbriae (Pili)

They are short and thin hair like filaments, formed of protein called pilin (antigenic). Fimbriae are responsible for attachment of bacteria to specific receptors of human cell (adherence). There are special types of pili called (sex pili) involved in conjunction.

Functions—Growth and Metabolism

Between successive cell divisions, cells grow through the functioning of cellular metabolism. Cell metabolism is the process by which individual cells process nutrient molecules. Metabolism has two distinct divisions: catabolism, in which the cell breaks down complex molecules to produce energy and reducing power, and anabolism, in which the cell uses energy and reducing power to construct complex molecules and perform other biological functions. Complex sugars consumed by the organism can be broken down into a less chemically complex sugar molecule called glucose. Once inside the cell, glucose is broken down to make adenosine triphosphate (ATP), a form of energy, through two different pathways.

The first pathway, glycolysis, requires no oxygen and is referred to as anaerobic metabolism. Each reaction is designed to produce some hydrogen ions that can then be used to make energy packets (ATP). In prokaryotes, glycolysis is the only method used for converting energy. The second pathway, called the Krebs cycle, or citric acid cycle, occurs inside the mitochondria and can generate enough ATP to run all the cell functions.

Within the nucleus of the cell (*light blue*), genes (DNA, *dark blue*) are transcribed into RNA. This RNA is then subject to post-transcriptional modification and control, resulting in a mature mRNA (*red*) that is then transported out of the nucleus and into the cytoplasm (*peach*), where it undergoes translation into a protein. mRNA is translated by ribosomes (*purple*) that match the three-base codons of the mRNA to the three-base anti-codons of the appropriate tRNA. Newly synthesized proteins (*black*) are often further modified, such as by binding to an effector molecule (*orange*), to become fully active.

Creation

Cell division involves a single cell (called a *mother cell*) dividing into two daughter cells. This leads to growth in multicellular organisms (the growth of tissue) and to procreation (vegetative reproduction) in unicellular organisms.

Prokaryotic cells divide by binary fission. Eukaryotic cells usually undergo a process of nuclear division, called mitosis, followed by division of the cell, called cytokinesis. A diploid cell may also undergo meiosis to produce haploid cells, usually four. Haploid cells serve as gametes in multicellular organisms, fusing to form new diploid cells.

DNA replication, or the process of duplicating a cell's genome, is required every time a cell divides. Replication, like all cellular activities, requires specialized proteins for carrying out the job.

Protein Synthesis

Cells are capable of synthesizing new proteins, which are essential for the modulation and maintenance of cellular activities. This process involves the formation of new protein molecules from amino acid building blocks based on information encoded in DNA/RNA. Protein synthesis generally consists of two major steps: transcription and translation.

Transcription is the process where genetic information in DNA is used to produce a complementary RNA strand. This RNA strand is then processed to give messenger RNA (mRNA), which is free to migrate through the cell. mRNA molecules bind to protein-RNA complexes called ribosomes located in the cytosol, where they are translated into polypeptide sequences. The ribosome mediates the formation of a polypeptide sequence based on the mRNA sequence. The mRNA sequence directly relates to the polypeptide sequence by binding to transfer RNA (tRNA) adapter molecules in binding pockets within the ribosome. The new polypeptide then folds into a functional three-dimensional protein molecule.

Movement or Motility

Cells can move during many processes: such as wound healing, the immune response and cancer metastasis. For wound healing to occur, white blood cells and cells that ingest bacteria move to the wound site to kill the microorganisms that cause infection.

At the same time fibroblasts (connective tissue cells) move there to remodel damaged structures. In the case of tumour development, cells from a primary tumour move away and spread to other parts of the body. Cell motility involves many receptors, crosslinking, bundling, binding, adhesion, motor and other proteins. The process is divided into three steps – protrusion of the leading edge of the cell, adhesion of the leading edge and de-adhesion at the cell body and rear, and cytoskeletal contraction to pull the cell forward. Each step is driven by physical forces generated by unique segments of the cytoskeleton.

Evolution

The origin of cells has to do with the origin of life, which began the history of life on Earth.

Origin of the First Cell

There are several theories about the origin of small molecules that could lead to life in an early Earth. One is that they came from meteorites. Another is that they were created at deep-sea vents. A third is that they were synthesized by lightning in a reducing atmosphere; although it is not clear if Earth had such an atmosphere. There are essentially no experimental data defining what the first self-replicating forms were. RNA is generally assumed to be the earliest self-replicating molecule, as it is capable of both storing genetic information and catalyzing chemical reactions. But some other entity with the potential to self-replicate could have preceded RNA, like clay or peptide nucleic acid.

Cells emerged at least 4.0–4.3 billion years ago. The current belief is that these cells were heterotrophs. An important characteristic of cells is the cell membrane, composed of a bilayer of lipids. The early cell membranes were probably more simple and permeable than modern ones, with only a single fatty acid chain per lipid. Lipids are known to spontaneously form bilayered vesicles in water, and could have preceded RNA. But the first cell membranes could also have been produced by catalytic RNA, or even have required structural proteins before they could form.

Origin of Eukaryotic Cells

The eukaryotic cell seems to have evolved from a symbiotic community of prokaryotic cells. DNA-bearing organelles like the mitochondria and the chloroplasts are almost certainly what remains of ancient symbiotic oxygen-breathing proteobacteria and cyanobacteria, respectively, where the rest of the cell seems to be derived from an ancestral archaean prokaryote cell – a theory termed the endosymbiotic theory.

There is still considerable debate about whether organelles like the hydrogenosome predated the origin of mitochondria, or viceversa: see the hydrogen hypothesis for the origin of eukaryotic cells.

Sex, as the stereotyped choreography of meiosis and syngamy that persists in nearly all extant eukaryotes, may have played a role in the transition from prokaryotes to eukaryotes. An 'origin of sex as vaccination' theory suggests that the eukaryote genome accreted from prokaryan parasite genomes in numerous rounds of lateral gene transfer. Sex-as-syngamy (fusion sex) arose when infected hosts began swapping nuclearized genomes containing co-evolved, vertically transmitted symbionts that conveyed protection against horizontal infection by more virulent symbionts.

History

- 1632–1723: Antonie van Leeuwenhoek teaches himself to grind lenses, builds a microscope and draws protozoa, such as *Vorticella* from rain water, and bacteria from his own mouth.
- 1665: Robert Hooke discovers cells in cork, then in living plant tissue using an early microscope.
- 1839: Theodor Schwann and Matthias Jakob Schleiden elucidate the principle that plants and animals are made of cells, concluding that cells are a common unit of structure and development, and thus founding the cell theory.
- The belief that life forms can occur spontaneously (*generatio spontanea*) is contradicted by Louis Pasteur (1822–1895) (although Francesco Redi had performed an experiment in 1668 that suggested the same conclusion).
- 1855: Rudolf Virchow states that cells always emerge from cell divisions (*omnis cellula ex cellula*).
- 1931: Ernst Ruska builds first transmission electron microscope (TEM) at the University of Berlin. By 1935, he has built an EM with twice the resolution of a light microscope, revealing previously unresolvable organelles.
- 1953: Watson and Crick made their first announcement on the double-helix structure for DNA on February 28.
- 1981: Lynn Margulis published *Symbiosis in Cell Evolution* detailing the endosymbiotic theory.

Carbohydrate

A carbohydrate is an organic compound which has the empirical formula $C_m(H_2O)_n$; that is, consists only of carbon, hydrogen and oxygen, with a hydrogen:oxygen atom ratio of 2:1 (as in water). Carbohydrates can be viewed as hydrates of carbon, hence their name. Structurally however, it is more accurate to view them as polyhydroxy aldehydes and ketones.

The term is most common in biochemistry, where it is a synonym of saccharide. The carbohydrates (saccharides) are divided into four chemical groupings: monosaccharides, disaccharides, oligosaccharides, and polysaccharides. In general, the monosaccharides and disaccharides, which are smaller (lower molecular weight) carbohydrates, are commonly referred to as sugars. While the scientific nomenclature of carbohydrates is complex, the names of the monosaccharides and disaccharides very often end in the suffix -ose. For example, blood sugar is the monosaccharide glucose, table sugar is the disaccharide sucrose, and milk sugar is the disaccharide lactose.

Carbohydrates perform numerous roles in living things. Polysaccharides serve for the storage of energy (e.g., starch and glycogen) and as structural components (e.g., cellulose in plants and chitin in arthropods). The 5-carbon monosaccharide ribose is an important component of coenzymes (e.g., ATP, FAD, and NAD) and the backbone of the genetic molecule known as RNA. The related deoxyribose is a component of DNA. Saccharides and their derivatives include many other important biomolecules that play key roles in the immune system, fertilization, preventing pathogenesis, blood clotting, and development.

In food science and in many informal contexts, the term carbohydrate often means any food that is particularly rich in the complex carbohydrate starch (such as cereals, bread and pasta) or simple carbohydrates, such as sugar (found in candy, jams and desserts).

Structure

Formerly the name "carbohydrate" was used in chemistry for any compound with the formula $C_m(H_2O)_n$. Following this definition, some chemists considered formaldehyde CH_2O to be the simplest carbohydrate, while others claimed that title for glycolaldehyde Today the term is generally understood in the biochemistry sense, which excludes compounds with only one or two carbons.

Natural saccharides are generally built of simple carbohydrates called monosaccharides with general formula $(CH_2O)_n$ where n is three or more. A typical monosaccharide has the structure $H\text{-}(CHOH)_x(C{=}O)\text{-}(CHOH)_y\text{-}H$, that is, an aldehyde or ketone with many hydroxyl groups added, usually one on each carbon atom that is not part of the aldehyde or ketone functional group. Examples of monosaccharides are glucose, fructose, and glyceraldehyde. However, some biological substances commonly called "monosaccharides" do not conform to this formula (e.g., uronic acids and deoxy-sugars such as fucose), and there are many chemicals that do conform to this formula but are not considered to be monosaccharides (e.g., formaldehyde CH_2O and inositol $(CH_2O)_6$).

The open-chain form of a monosaccharide often coexists with a closed ring form where the aldehyde/ketone carbonyl group carbon (C=O) and hydroxyl group (-OH) react forming a hemiacetal with a new C-O-C bridge. Monosaccharides can be linked together into what are called polysaccharides (or oligosaccharides) in a large variety of ways. Many carbohydrates contain one or more modified monosaccharide units that have had one or more groups replaced or removed. For example, deoxyribose, a component of DNA, is a modified version of ribose; chitin is composed of repeating units of N-acetylglucosamine, a nitrogen-containing form of glucose.

Monosaccharides

Classification of Monosaccharides

Monosaccharides are classified according to three different characteristics: the placement of its carbonyl group, the number of carbon atoms it contains, and its chiral handedness. If the carbonyl group is an aldehyde, the monosaccharide is an aldose; if the carbonyl group is a ketone, the monosaccharide is a ketose. Monosaccharides with three carbon atoms are called trioses, those with four are called tetroses, five are called pentoses, six are hexoses, and so on. These two systems of classification are often combined. For example, glucose is an aldohexose (a six-carbon aldehyde), ribose is an aldopentose (a five-carbon aldehyde), and fructose is a ketohexose (a six-carbon ketone).

Each carbon atom bearing a hydroxyl group (-OH), with the exception of the first and last carbons, are asymmetric, making them stereocenters with two possible configurations each (R or S). Because of this asymmetry, a number of isomers may exist for any given monosaccharide formula. The aldohexose D-glucose, for example, has the formula $(C\cdot H_2O)_6$, of which all but two of its six carbons atoms are stereogenic, making D-glucose one of $2^4 = 16$ possible stereoisomers. In the case of glyceraldehyde, an aldotriose, there is one pair of possible stereoisomers, which are enantiomers and epimers. 1,3-dihydroxyacetone, the ketose corresponding to the aldose glyceraldehyde, is a symmetric molecule with no stereocenters). The assignment of D or L is made according to the orientation of the asymmetric carbon furthest from the carbonyl group: in a standard Fischer projection if the hydroxyl group is on the right the molecule is a D sugar, otherwise it is an L sugar. The "D-" and "L-" prefixes should not be confused with "d-" or "l-", which indicate the direction that the sugar rotates plane polarized light. This usage of "d-" and "l-" is no longer followed in carbohydrate chemistry.

Ring-straight Chain Isomerism

The aldehyde or ketone group of a straight-chain monosaccharide will react reversibly with a hydroxyl group on a different carbon atom to form a hemiacetal or hemiketal, forming a heterocyclic ring with an oxygen bridge between two carbon atoms. Rings with five and six atoms are called furanose and pyranose forms, respectively, and exist in equilibrium with the straight-chain form.

During the conversion from straight-chain form to the cyclic form, the carbon atom containing the carbonyl oxygen, called the anomeric carbon, becomes a stereogenic centre with two possible configurations: The oxygen atom may take a position either above or below the plane of the ring. The resulting possible pair of stereoisomers are called anomers. In the α *anomer*, the -OH substituent on the anomeric carbon rests on the opposite side (trans) of the ring from the CH_2OH side branch. The alternative form, in which the CH_2OH substituent and the anomeric hydroxyl are on the same side (cis) of the plane of the ring, is called the β *anomer*. You can remember that the β anomer is cis by the mnemonic, "It's always better to βe up". Because the ring and straight-chain forms readily interconvert, both anomers exist in equilibrium. In a Fischer Projection, the α anomer is represented with the anomeric hydroxyl group *trans* to the CH_2OH and *cis* in the β anomer.

Use in Living Organisms

Monosaccharides are the major source of fuel for metabolism, being used both as an energy source (glucose being the most important in nature) and in biosynthesis. When monosaccharides are not immediately needed by many cells they are often converted to more space efficient forms, often polysaccharides. In many animals, including humans, this storage form is glycogen, especially in liver and muscle cells. In plants, starch is used for the same purpose.

Disaccharides

Two joined monosaccharides are called a disaccharide and these are the simplest polysaccharides. Examples include sucrose and lactose. They are composed of two monosaccharide units bound together by a covalent bond known as a glycosidic linkage formed via a dehydration reaction, resulting in the loss of a hydrogen atom from one monosaccharide and a hydroxyl group from the other.

The formula of unmodified disaccharides is $C_{12}H_{22}O_{11}$. Although there are numerous kinds of disaccharides, a handful of disaccharides are particularly notable.

Sucrose, pictured to the right, is the most abundant disaccharide, and the main form in which carbohydrates are transported in plants. It is composed of one D-glucose molecule and one D-fructose molecule. The systematic name for sucrose, *O*-α-D-glucopyranosyl-(1→2)-D-fructofuranoside, indicates four things:

- Its monosaccharides: glucose and fructose
- Their ring types: glucose is a pyranose, and fructose is a furanose

- How they are linked together: the oxygen on carbon number 1 (C1) of α-D-glucose is linked to the C2 of D-fructose.
- The *-oside* suffix indicates that the anomeric carbon of both monosaccharides participates in the glycosidic bond.

Lactose, a disaccharide composed of one D-galactose molecule and one D-glucose molecule, occurs naturally in mammalian milk. The systematic name for lactose is *O*-β-D-galactopyranosyl-(1→4)-D-glucopyranose. Other notable disaccharides include maltose (two D-glucoses linked α-1,4) and cellulobiose (two D-glucoses linked β-1,4). disaccharides can be classified into two types they are reducing and non-reducing disaccahrides if the functional group is present in bonding with another sugar unit it is called as reducing disaccharide.

Oligosaccharides and Polysaccharides

Oligosaccharides and polysaccharides are composed of longer chains of monosaccharide units bound together by glycosidic bonds. The distinction between the two is based upon the number of monosaccharide units present in the chain. Oligosaccharides typically contain between three and ten monosaccharide units, and polysaccharides contain greater than ten monosaccharide units. Definitions of how large a carbohydrate must be to fall into each category vary according to personal opinion. Examples of oligosaccharides include the disaccharides mentioned above, the trisaccharide raffinose and the tetrasaccharide stachyose.

Oligosaccharides are found as a common form of protein posttranslational modification. Such posttranslational modifications include the Lewis and ABO oligosaccharides responsible for blood group classifications and so of tissue incompatibilities, the alpha-Gal epitope responsible for hyperacute rejection in xenotransplantation, and O-GlcNAc modifications.

Polysaccharides represent an important class of biological polymers. Their function in living organisms is usually either structure- or storage-related. Starch (a polymer of glucose) is used as a storage polysaccharide in plants, being found in the form of both amylose and the branched amylopectin. In animals, the structurally similar glucose polymer is the more densely branched glycogen, sometimes called 'animal starch'. Glycogen's properties allow it to be metabolized more quickly, which suits the active lives of moving animals.

Cellulose and chitin are examples of structural polysaccharides. Cellulose is used in the cell walls of plants and other organisms, and is claimed to be the most abundant organic molecule on earth. It has many uses such as a significant role in the paper and textile industries, and is used as a feedstock for the production of rayon (via the viscose process), cellulose acetate, celluloid, and nitrocellulose. Chitin has a similar structure, but has nitrogen-containing side branches, increasing its strength. It is found in arthropod exoskeletons and in the cell walls of some fungi. It also has multiple uses, including surgical threads. Other polysaccharides include callose or laminarin, chrysolaminarin, xylan, arabinoxylan, mannan, fucoidan and galactomannan.

Nutrition

Foods high in carbohydrate include fruits, sweets, soft drinks, breads, pastas, beans, potatoes, bran, rice, and cereals. Carbohydrates are a common source of energy in living organisms, however, no carbohydrate is an essential nutrient in humans. Carbohydrates are not necessary building blocks of other molecules, and the body can obtain all its energy from protein and fats. The brain and neurons generally cannot burn fat for energy, but use glucose or ketones. Humans can synthesize some glucose (in a set of processes known as gluconeogenesis from specific amino acids, from the glycerol backbone in triglycerides and in some cases from fatty acids. Carbohydrate contains 15.8 kilojoules (3.75 kilocalories) and proteins 16.8 kilojoules (4 kilocalories) per gram, while fats contain 37.8 kilojoules (9 kilocalories) per gram. In the case of protein, this is somewhat misleading as only some amino acids are usable for fuel. Organisms typically cannot metabolize all types of carbohydrate to yield energy. Glucose is a nearly universal and accessible source of calories. Many organisms also have the ability to metabolize other monosaccharides and Disaccharides, though glucose is preferred. In *Escherichia coli*, for example, the *lac* operon will express enzymes for the digestion of lactose when it is present, but if both lactose and glucose are present the *lac* operon is repressed, resulting in the glucose being used first. Polysaccharides are also common sources of energy. Many organisms can easily break down starches into glucose, however, most organisms cannot metabolize cellulose or other polysaccharides like chitin and arabinoxylans.

These carbohydrates types can be metabolized by some bacteria and protists. Ruminants and termites, for example, use microorganisms to process cellulose. Even though these complex carbohydrates are not very digestible, they may comprise important dietary elements for humans. Called dietary fibre, these carbohydrates enhance digestion among other benefits.

Based on the effects on risk of heart disease and obesity, the Institute of Medicine recommends that American and Canadian adults should get between 45–65% of dietary energy from carbohydrates. The Food and Agriculture Organization and World Health Organization jointly recommend that national dietary guidelines set a goal of 55–75% of total energy from carbohydrates, but only 10% directly from sugars (their term for simple carbohydrates).

Classification

Historically nutritionists have classified carbohydrates as either simple or complex, however, the exact delineation of these categories is ambiguous.

Today, simple carbohydrate typically refers to (monosaccharides and disaccharides) and complex to (oligosaccharides and polysaccharides). However, the term *complex carbohydrate* was first used in slightly different context in the U.S. Senate Select Committee on Nutrition and Human Needs publication *Dietary Goals for the United States* (1977). In this work, complex carbohydrate were defined as "fruit, vegetables and whole-grains". Some nutritionists still appear to use complex carbohydrate to refer to any sort of digestible saccharide present in a whole food, where fibre, vitamins and minerals are also found (as opposed to processed carbohydrates which provide calories but few other nutrients).

A commonly held belief, even among nutritionists, is that complex carbohydrates (polysaccharides, e.g. starches) are digested more slowly than simple carbohydrates (sugars) and thus are healthier.

However, there appears to be no significant difference between simple and complex carbs in terms of their effect on blood sugar Some simple carbohydrates (e.g. fructose) are digested very slowly, while some complex carbohydrates (starches), especially if processed, raise blood sugar rapidly. The speed of digestion is determined by a variety of factors including which other nutrients are consumed with the carbohydrate, how the food is prepared, individual differences in metabolism and the chemistry of the carbohydrate.

Dietary guidelines generally recommend that complex carbohydrates (starches) and nutrient-rich simple carbohydrates such as fruits and vegetables, and dairy products make up the bulk of carbohydrate consumption. Highly processed sources of carbohydrate such as corn or potato chips, candy, sugary drinks, pastries and white rice are generally considered unhealthy in excess.

The USDA's *Dietary Guidelines for Americans 2005* dispensed with the simple/complex distinction, instead recommending fibre-rich foods and whole grains. The glycemic index and glycemic load concepts have been developed to characterize food Behaviour during human digestion. They rank carbohydrate-rich foods based on the rapidity and magnitude of their effect on blood glucose levels. Glycemic index is a measure of how quickly food glucose is absorbed, while glycemic load is a measure of the total absorbable glucose in foods.

The insulin index is a similar, more recent classification method that ranks foods based on their effects on blood insulin levels, which are caused by glucose (or starch) and some amino acids in food.

10

Heredity in Animals

Frogs grow old as well as men, and will die of old age even if they have not previously met death by some accident. If there is to be a race of frogs, there must therefore be a continual production of new frogs to take the place of the old. Our next inquiry must be into the method adopted to ensure this *reproduction*. If we run an individual frog's history backwards through our minds, as a film can be run backwards through a cinema, we find that the full-sized or adult frog was preceded by a young frog of much smaller size but of the same general shape and structure.

Before this, however, the frog was so different as to merit a different name; it was a tadpole, lived only in the water, fed mainly on vegetable food, had only rudimentary legs, and swam with a tail. It did not breathe by lungs, but by gills; these are branched outgrowths of the body-wall with many capillaries just below a very thin skin.

They grow on the outer borders of slits which lead from the cavity of the mouth to the side of the neck, and, since water is continually being sucked in at the mouth and then forced out through these gill-slits, they can be always taking up dissolved oxygen from the water and passing carbon dioxide out into it. In large and mediumsized tadpoles, the gills and gill-slits are covered by a flap of skin or gill-cover; but in quite small ones the gills stand out free on the side of the neck.

As the adult frog grew out of a smaller frog, so the large tadpole has grown from a smaller tadpole of the same general form and containing the same systems of organs, except that it is altogether without limbs. The small tadpole, in its turn, hatched out of a glutinous covering, which with several hundred others formed a mass of frog-spawn. If we had gone farther back and examined the developing frog inside the jelly a few days before hatching, we should have seen that it still had the general form of a tadpole; but its gills were mere knobs, its tail not fully formed.

Figure: *Tadpoles*

Before that again there was a time when, although the general plan of the internal organs was the same, the organs themselves were not present as working pieces of machinery, but merely blocked out in an undifferentiated state, with their component cells not yet specialized as in the adult.

The liver at this stage, for instance, was represented merely by an unbranched tube; the heart was an S-shaped blood-vessel, as yet without muscles or valves; the chief parts of brain and spinal cord could have been seen, but in a much more rudimentary shape than in the full-grown frog, and with no nerve-fibres yet formed by outgrowth from their cells.

Still earlier, the developing animal had no particular resemblance to a tadpole, and even the main organ-systems were barely recognizable. The future brain and spinal cord, for example, were represented by a groove along the back; the digestive system was an irregular space within a mass of yolk in the interior. Before this, the organism (when passing through what is termed the gastrula stage) was quite spherical, consisting of an outer sheet of several layers of small cells, an inner sheet mainly composed of big yolk-laden cells surrounding the cavity which is the first rudiment of the gut, and a middle sheet between the other two. A day or so earlier, in the blastula stage, there was but one sheet of cells, and no digestive rudiment. Earlier still, there were only a few large cells; and at the last, we can trace our frog back to a single very large cell, black above and white below, loaded with reserve food in the shape of yolk, and containing but a single nucleus. This is the fertilized egg. 'What is the origin of the fertilized egg?' is the next obvious question. But before attempting to answer this, we must go in somewhat more detail into the actual progress of development. This can be divided into a number of periods. The egg contains plenty of yolk, in order to provide food for the young frog before it can feed for itself. It is consequently a very large cell, and the first step to be taken is to divide it into cells of a more convenient size—the bricks to be used in the future building. The egg divides into two, these into four, and so on, until the blastula stage is reached. This marks the close of the first period, usually called the period of segmentation. The second period ends in the formation of a rough—a very rough—ground-plan of the future organism. A fold of the dark, smaller cells grows over the larger yolky cells, and the crack between the inner layer of the fold and the yolk afterwards swells out to be the first rudiment of the digestive system. If we hold one end of a sheet of paper in place and bring the other end down towards the first, we will produce a similar fold; but the movement of the living fold is caused by rapid growth of cells just about the folded part. Between the inner and outer layers of the fold a third layer of cells is split off.

Now the first groundplan is ready—there are three distinct layers of cells in existence. The outer layer will later give rise to the epidermis of the skin, the sense-organs, and the nervous system; the inner layer to the gut and all its appendages such as liver, pancreas, thyroid, lungs, and gill-slits; and the middle layer to the muscles, the skeleton, the connective tissue, the bloodsystem, the kidneys, and the reproductive organs. These three primary layers are called germ-layers, so that this is the period of germ-layer formation. Segmentation and germlayer formation are a good deal simpler in forms with less yolk in their eggs. In the third period a great advance is made—the main systems of organs are blocked out. The embryo lengthens: a groove forms on the back, deepens, and its sides arch over and meet to produce a tube; this tube is the rudiment of the whole central nervous system, and of most of the nerves. Tiny pits appear on the side of the head, representing the future nose and ear; and from near the front of the nervetube, two hollow outgrowths arise which will form the main part of the eyes. Below the nerve-tube, a long straight rod is nipped off from the top of the gut.

This is the notochord, the early and less complicated precursor of the true backbone. On each side of the notochord, the middle layer of cells grows rapidly, and cuts itself up into a series of blocks of tissue, the muscle-segments; from these the voluntary muscles will be formed. Below them, a split appears in the middle layer—the rudiment of the body-cavity. From its walls a series of funnel-shaped tubes grow out, and their ends unite and form a duct which grows back to open at the hind-end—the rudiments of kidneys and of ureters.

The mouth and anus are pierced; outgrowths of the gut produce the pancreas- and liver-rudiments; other outgrowths reach the exterior and will form the gill-slits. Scattered cells of the middle layer unite to form tubes, and these tubes join up to give the blood-vessels. The main characteristic of the period has been the formation, in a very short time (not more than 48 hours at ordinary temperatures), of a great deal of visible structure where very little was to be seen before.

We may call this the period of primary differentiation of organs. At the end of this period, the laying out of the detailed ground-plan, some little time before hatching, the chief organ-systems and organs are thus all present, but they are not yet in working order—their cells are still all of the same general type, not specialized for their various duties. The next, or fourth period, therefore, is primarily chiefly that of the differentiation of the tissues. The gut-rudiment, growing capable of digesting, becomes a gut; the muscle-rudiments become muscles; the nerve-tube becomes a brain, spinal cord, and nerves; and so on with the rest of the organs of the body.

This change is accomplished by the time of hatching for some organs, a little later for others. It takes place through transformation of the shape and structure of cells. In the next period the animal *works*, and becomes selfsupporting; feeding and consequent growth are its chief characteristics; but towards its close, preparation is made for the adult stage by the appearance of limbs and of lungs-again first as mere rudiments, later as organs capable of working.

This is the *larval* period, a larva being a stage in an animal's history when it is self-supporting but (and therefore no longer an embryo) radically different in structure and mode of life from the adult. The sixth stage, or period of metamorphosis, is a violent transition from larva to adult, from water to land.

If we could not actually follow the transformation, it would be impossible to guess that a young frog and a tadpole were stages in one and the same animal's life. The limbs grow rapidly, the shape of head and trunk and the colour of the skin become altered, and there are internal alterations, such as a shortening of the intestine in view of the change from vegetable to animal diet, the remodelling of the skull, and the replacement of much cartilage by bone. But the most remarkable changes are those affecting the tail and gills.

These do not drop off, as is still believed by some unobservant people; they are absorbed. They have been built up; now they are unbuilt. Their tissues lose their differentiation, degenerate, and are used as food-material. Thus the development of an organ need not always be forwards; it may be reversed. After the change, another growth-stage sets in, which we may call the juvenile period, when the animal is definitely a frog, but not yet grown up. After a time, however, growth slows down, and at about the same time sexual maturity begins. So the eighth of our periods, the adult period, starts. This is the longest of all the periods, and is best developed in the highest forms of life. It is in them a condition in which neither growth nor breakdown has the upper hand, a period of balanced activity, which has been compared to the apparent rest of a 'sleeping' top. Frogs, like mammals and birds, have a well-marked adult period in which growth is absent. Many lower animals, however, such as most crabs and shell-fish, go on growing throughout life, and in them sexual maturity, not cessation of growth, is the only criterion of the adult phase. Finally, however, the ninth period sets in—old age. The metabolism grows feebler, the organs no longer work so well, and even if a violent end does not terminate the animal's existence, as is almost always the case in nature, its life eventually comes to death as an inevitable close. Such a death is unavoidable and normal; we may call it natural death—a real wearing-out of the tissues, and a consequent collapse of the organism like the collapse of the hundredyear-old 'one-hoss shay' in the poem.

The development of an animal like the frog, then, consists partly of growth; partly of differentiation or increase of complexity; thirdly, of metamorphosis from one form to another; fourthly, of the attainment of a relatively stable, balanced state, in which growth and differentiation are almost absent; and lastly, of the loss of this stability, the wearing-out of the machine—old age, and death. Now having traced the development of a grown-up frog from a single fertilized egg, we can return to our question as to the origin of this fertilized egg. If we look at the ovaries or reproductive organs of a female frog in early spring, we shall find in them, and actually produced by them, a great number—one to two thousand—of eggs *apparently* similar to fertilized eggs; a little later in the season, these will be found loose in the swollen lower part of the oviducts, each surrounded with a thin layer of jelly. But if we take these eggs and put them in water, they will not develop.

For them to start development, fertilization is normally necessary; and by fertilization we mean union of the egg with a reproductive cell of the male. In all animals, the eggs are cells detached from these special organs of the body, the ovaries. The reproductive organs of the male are the testes, small whitish bodies consisting mainly of a number of microscopic tubes. The cells forming the walls of these tubes divide, and some of the fresh cells thus produced change their shape in a remarkable way; the nucleus becomes long and dense, and the rest of the cell becomes transformed into a sharp point at one end and a swimming tail at the other. They are now ripe, and are called spermatozoa or sperms. These are shed into the hollow of the tubes, pass down canals into and through the kidney, and down the ureter, to be stored in a pouch on its side.

A drop of the contents of this taken in early April and mixed with water will show thousands of sperms in active movement; a male frog produces billions in a season. When the female lays her eggs, the male sheds his sperm over them. The sperms are attracted by the eggs, and start burrowing into them. Once one has succeeded in forcing its nucleus inside, a change takes place over the egg's surface which prevents any further spermatozoa from entering. The sperm-nucleus then swells up, and becomes very like the nucleus of the egg. The two nuclei meet, and actually unite to form one. This completes the act of fertilization; and only after this does development begin. Sexual reproduction in the frog then consists in this: that union of two cells, or at least of the nuclei of two cells, takes place, that the cells at the moment of union are detached and independent, but previously formed part of the bodies of other individuals of the species; and that the cell formed by their union develops into a new individual.

It has been found possible with the eggs of frogs, worms, sea-urchins, and other animals, to make the egg start its development by artificial means (by chemical treatment in sea-urchins; by heat in starfish; by pricking the egg with a fine needle dipped in blood, in frogs), without any sperm being present, so producing fatherless animals.

Some of these have been raised to maturity, and appear perfectly normal. Thus fertilization consists really of two separate and separable processes—activation, or the starting-off of the egg on development, and the union of the nuclei of male and female reproductive cells. The two cells which thus unite are called *gametes* or 'marrying cells'; and the product of their union is called a *zygote*—something formed by the 'yoking together' of two gametes. Usually the gametes are, as in the frog, of two markedly different kinds, the female gamete large and yolk-laden, the male gamete tiny and active. This difference between the two gametes, however, is not universal.

In many low forms of life, both plant and animal, the two are alike; and in some, as in the single-celled animal Paramecium, only nuclei and not whole cells fuse with each other. Thus the essential fact in sexual fusion is the union of the two nuclei. The difference between the two gametes is only secondary; the size and yolk-contents of the egg serve to give the developing embryo a good start in life after fertilization; and the shape and activity of the sperm ensure that the two gametes shall meet.

The race of frogs, then, can be thought of as consisting of a number of continuous streams of living substance. The streams flow for the most part in the form of zygotes, starting as fertilized eggs and developing into mature frogs; but for some of the time they subdivide to flow in the form of gametes, and then such of these streams as do not die out, unite in pairs to become zygotes once more. Or, to put it in another way, the species *Rana temporaria*, or any other kind of higher animal, such as the human species, comprises not *two* but *four* kinds of individuals—male and female zygotes, and male and female gametes. The zygotes are large and long-lived, the gametes small, short-lived, but far more numerous.

To understand fully what this implies, we must again turn back. We have said that cells divide, but have not described the process. As a matter of fact, it is a very remarkable one. When a cell is ready to divide, the wall of the nucleus breaks down, and its contents change from their chiefly fluid state into a number of more solid rod- or strap-shaped bodies, which, because they become coloured by many dyes, are called *chromosomes*. Meanwhile, a starshaped figure of radiating fibres is seen in the cell. This divides into two, forming a spindle-shaped set of fibres with a radiating 'star' at each end, and the chromosomes arrange themselves where the fibres from the two stars meet, in the centre of the spindle.

The chromosomes next split right down their length, and the halves travel away from each other, one to one pole, the other to the other. Finally, the body of the cell divides, the set of chromosomes at either pole swells up to form a nucleus again, and there are two cells, each with its nucleus, in place of one. This process of nuclear division ensures that each chromosome is divided exactly along its length.

The units which determine the hereditary characters of an animal probably lie along the chromosomes, so that as a result of mitosis a complete set of halved units is distributed to both of the two cells arising at any division; and these halved units, since they are alive, soon all grow up to their original size again. The number of chromosomes is always the same for a given race of animals or plants; not only this, but often the individual chromosomes can be distinguished from each other by differences of size and shape; and when this is so, they can be arranged in a series of pairs, so that the total number is always even.

At one of the cell-divisions just before the formation of gametes, however, the process is quite different. Instead of the chromosomes dividing, the members of a pair come to lie side by side; and at division one whole chromosome of a pair is separated from the other. This process is called the *reduction* of the chromosomes, for owing to it, each of the two cells produced at this division possess only half the ordinary number of chromosomes for the species.

Accordingly, gametes have half the number of chromosomes found in zygotes; and the reason the chromosomes of the zygotes exist in pairs is that one member of the pair has come from the father and one from the mother.

Thus, the ordinary higher animal or plant has two complete sets of chromosomes, like two packs of cards.

This is of great importance in the study of heredity, or, in other words, of the way in which characteristics are handed on from one generation to the next. The first maturation division is also the reduction division, the $2n$ or diploid number of chromosomes found in the bodycells being reduced to n, the haploid number, through the separation of corresponding whole chromosomes from each other.

Here, in order to retain the large size of the egg, three of every four gametes produced are minute and non-functional, and are called polar bodies, while only one becomes a functional ovum. Note that there are two of each kind of chromosome. A frog gives rise to new frogs, not to toads or lizards; within a single species, such as man, many characters 'run in families'—physical characters like black hair, or mental characters like musical talent; and it is a common observation that children inherit traits from their parents and from remoter ancestors.

The facts of heredity are often obvious enough. But what is the machinery by which they are brought about? How is it that the tiny, simple-looking egg has the power of developing into the complicated adult, and an adult just of one particular kind? How is it possible that of two eggs, which would be practically indistinguishable under the microscope, one is predestined to produce a redhaired tall boy, the other a dark-haired medium-sized girl?

There is a great deal of evidence to support the idea that inheritance—the transmission of qualities or characters from one generation to the next—depends upon the actual transmission of small units of living substance in the gametes.

These units are called the *factors of heredity*, or sometimes still more shortly the *genes*; and there is further evidence that these factors are contained in the chromosomes.

Certain results flow from this—results, for instance, concerning the numerical proportions of different types to be expected on crossing different strains; and, indeed, all the subject-matter of the comparatively new science called Mendelism, after the Austrian Abbot Mendel, who first discovered and interpreted the essential facts.

Here we are concerned only with some of the more general aspects of the question. A sexually reproducing animal or plant grows up from a fertilized egg into its adult condition. The fact that it grows up in its own particular way, that a newt's egg will produce a newt and a frog's egg a frog, although they are scarcely distinguishable and although they grow up in the same pond, must depend on the eggs containing something —let us call it the hereditary constitution—which determines the way it shall grow up. Mendel discovered two important laws concerning the way in which the hereditary constitution of an animal (or plant) is related to that of its parents.

The first is usually called the *law of segregation*. This implies that the hereditary constitution is composed of a number of self-reproducing units, presumably of a chemical nature, the *factors* or *genes*. Normally there are two of each kind of unit in the body; but before gamete-formation, the two members of such a pair segregate, or separate from each other, so that each gamete has one member only of each kind of unit.

The second law is that of *independent recombination* of units. It sums up the fact, elicited by breeding-tests, that the segregation of members of different kinds of units is usually quite independent. It is easy to see that this is exactly what would be expected if the units were contained within the chromosomes; and, as a matter of fact, this has now been proved to be the case. The adult organism, whether human being, fly, rabbit, or pea-plant, has two complete sets of chromosomes. The two members of a pair separate from each other at reduction, so that each gamete receives one complete set; recent observations have further shown that the way in which one chromosome-pair separates at reduction is quite independent of other pairs.

Thus, the single set of chromosomes in one gamete need not all have come in from the father, or all from the mother; there must merely be one of each kind of chromosome present. If we compare the chromosomes to cards, fertilization is like the mixing-up of two packs of cards with different coloured backs. Reduction ensures that this double pack shall be sorted out into two single packs again, but pays no attention to the colours of the cards' backs. Each sorted-out pack will be complete in having one card of each kind, but any of the cards may have either a red or a blue back. Thus it will be most unlikely that, as regards the precise combination of red and blue backs, any of the sorted-out packs will be identical with any other, since the number of possible combinations is so enormous. Each chromosome appears to contain a large number of units, and the segregation of two units will only be quite independent when they lie in different chromosomes.

When they are both in the same chromosome they tend to stick together more often than they separate. The units are the really important things in the hereditary constitution, and the chromosomes appear to be merely convenient lengths, so to speak, into which the hereditary constitution is cut up. Accordingly, the cards in our simile should really be taken to represent the separate units, and not the compound units we call chromosomes, which would really be more like suits.

If we do this, however, we must add one more point to make our analogy complete. We must suppose that any particular card need not always have precisely the same value, but that we could have one ten of hearts, say, that was a little above par, and another a little below par. In terms of factors, this would mean that although a given kind of factor always exerted the same general kind of effect in development, yet it might exist under different forms, with slight differences in effect. For instance, we know that one factor concerned with colour-production in rabbits may exist in at least four forms, one producing no pigment at all, as in albinos; the next allowing colour to appear in the ears and muzzle and other extremities, as in the Himalayan breed; the next giving the Chinchilla coat; and the last allowing full development of pigment.

The different forms of one factor are called *allelomorphs*. So far we have dealt with the generalities of Mendelian theory. They will become much clearer if we look at a couple of examples. If a certain breed of black fowl is crossed with another strain which is white with a few black markings, the offspring will all be unlike either parent, of a type known as the Blue Andalusian, which is a bluish or diluted black colour, with white 'lacing' on the feathers.

Poultry-fanciers have long tried to get a pure breed of these Andalusians by mating them with each other, but always without success—the birds 'threw' blacks and whites as well as producing more blues. When the subject came to be accurately worked out it was found that when Andalusians are mated with each other, on an average 25 per cent of their offspring are blacks, 25 per cent. whites, and 50 per cent.

Andalusian again. Not only this, but the blacks bred true when crossed with each other, and so did the whites, but the Andalusians always gave the same proportions of blacks, whites, and Andalusians once more. The organisms are represented as large circles, the genes as small circles within them. In the gametes, only the genes are represented. Unshaded represents splashed-white; black represents the gene for black, and close dotting visible black; sparse dotting represents visible blue. When a black cock is crossed with a white hen all his gametes (the sperms) contain one factor for black, all hers (the unfertilized eggs) one factor for white.

The offspring will have one of either factor; and their interaction will bring about the so-called blue colour. Exactly the same result will happen if a white cock is crossed with a black hen. According to Mendel's second law, factors for different characters are inherited independently of each other. For instance, to take fowls once more, the ordinary type of comb, called single, depends on one form of a particular factor; that called rose, which is low and long, upon another. When a cross is made between birds with these two sorts of comb, the hybrid is indistinguishable in appearance from the pure 'rose' stock.

This introduces us to the fact that one form of a factor may mask another, and be what is called *dominant* to it. But the factor for single-comb is still present in the hybrids, as is shown by breeding-tests. For whereas the original pure-bred rose-comb stock gives nothing but rose-combs, the hybrids crossed with each other give 25 per cent. pure-breeding singles, 25 per cent. pure-breeding rose-combs, and 50 per cent. hybrid rose-comb birds that again do not breed true.

That is to say, the difference between single- and rose-comb depends upon a difference between two forms of the same hereditary factor. Now, if we cross a black bird with a single-comb and a white bird with a rose-comb all the offspring will clearly be blue Andalusian in colour, with rose-combs.

But we might expect either of two things to happen in the second generation. Either the factors for black and single will stick together when the first generation forms its gametes; and then the only pure-breeding types we shall get will be black single-combs and white rose-combs again, like the grandparents. Or else the factors can separate; if so, then we shall get gametes containing the factor for white with that for single-comb, and those containing factors for black and for rose-comb, as well as the other two types.

In this case, in addition to types which will not breed true, we should get four pure-breeding types in the second generation—not only blacks with single-combs and whites with rose-combs like the grandparents, but also two new types, blacks with rose-combs and whites with single-combs. The fact of the independent segregation of different pairs of factors is of the greatest practical as well as theoretical importance, for it enables us to take to pieces the hereditary constitution, so to speak, and put it together again in new ways.

More definitely, it enables us to take one particular Mendelian character which happens to be desirable in an otherwise undesirable strain or breed, and to combine it with other desirable characters in another breed. A great deal of Mendelian work has been done with the tiny fruit-fly Drosophila, of which over ten million have been bred by Morgan in America in pedigreed cultures.

Not all factor-differences produce such big or such clearcut differences in appearance as we have been considering. Sometimes the effects are very small, or the effect of one factor interacts with that of another, and the analysis becomes correspondingly more difficult.

But the principles discovered by Mendel remain the same—the hereditary constitution consists of separate units, and these segregate at reduction. This discovery is on the biological plane somewhat comparable to the discovery of the atomic structure of matter. One point concerning human heredity deserves mention. People are often puzzled at the great diversity of physique and temperament and ability between members of the same family, although born of the same parents and brought up in the same home.

This, however, was only difficult to understand before we had Mendel's principles to guide us. For such diversity is exactly what is to be prophesied if there is independent segregation of self-perpetuating factors. It is extremely unlikely that two human gametes, even from the same parents, will receive an identical outfit of factors at reduction; and even if the mother were to achieve the improbable result of producing two identically constituted ova, and the father two sperms whose constitutions were alike, there would again be enormous odds (many million to one) against the two similar sperms, among their huge crowd of competitors, happening to fertilize the two similar eggs. On the other hand, if a fertilized ovum were to divide into two, and each part developed into a whole organism, we should expect both to be very similar, since both would possess the same hereditary chromosome outfit. This apparently is what happens in the production of so-called 'identical' twins, which as we know are often so alike as to be indistinguishable. Another difficulty often raised is the fact that genius usually is not inherited. Again, however, this is to be expected on Mendelian principles. If, as is probable, genius is the result of a very unusual hereditary combination of factors, then at the formation of gametes this combination will be inevitably taken to pieces again.

It should not be forgotten, however, that when we take the averages of large numbers these irregularities become smoothed out, and we find a strong average resemblance, due to heredity, between parent and offspring, or between brothers and sisters. The science of biometry deals with studies of this sort, and is very important when we are dealing with heredity in human beings, as in them, of course, experimental breeding is impossible, and the collection and handling of statistics is often the best method available. It appears, then, that the hereditary factors not only are contained in the chromosomes, but that each one is always situated in a particular chromosome, and in a particular spot in that chromosome.

Thus, each chromosome is different from every other, and each contains a great number of these factors arranged along it in a definite order, each factor always keeping between the same two neighbour-factors. Further, it appears that by far the greater part of the factors that determine inheritance from one generation to the next are situated in this way in the chromosomes. Thus, the chromosomes are, or at least contain, the hereditary constitution of the animal or plant, and all the chromosomes taken together, from this point of view, constitute one great unit, composed of a very large number of smaller units—the factors—which are arranged within the large unit in definite positions, and in definite proportions.

The whole is like a gigantic single organic chemical compound, since the molecules of such a chemical compound are all made up of smaller parts—the side-chains and radicals and single atoms. Here, however, each of the constituent units is alive, and must have the power of self-reproduction; thus, the linking of the units together in a definite way ensures the constancy of the whole.

If there had been an excess multiplication of one of the factors during the resting period of a cell, even so, when the cell came to divide and the chromosomes to form, only the same relative amount of the factor could become linked up with the other factors of its chromosome. So the chromosomes are not only a selfmultiplying but also a self-regulating piece of machinery.

In ordinary inheritance, then, we have the handing down from parent to offspring of chromosome outfits, or, in other words, of sets of hereditary factors; and it is on these that the constancy of a species depends. They decide that a fowl's egg shall become a fowl, and a duck's egg a duck, even though they are hatched in the same incubator. They, reacting with the environment, determine what the adult organism shall be. Apparently, however, slight changes sometimes occur in the factors, usually in a single one at one time: these changes are called *mutations*.

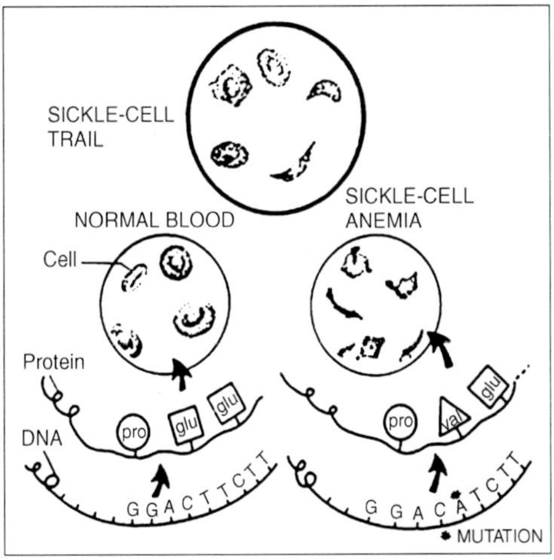

Figure: Mutations

Whereas an animal may normally have red eyes or a grey body-colour, mutations may arise which cause it, even though brought up in the same conditions, to have pink eyes or a black body-colour. When mutations occur, they are inherited in Mendelian fashion; the separation of pairs of chromosomes at reduction, and the subsequent meeting of various sets at random in fertilization, shuffles and recombines in every possible way any mutations that do occur.

Thus, while the chromosomes in general act as a regulator of the constancy of the race, the occurrence of mutations and their subsequent combination in new ways by sexual reproduction provide the possibility of change. The idea of separate factors is important, for it means that when we are dealing with an animal's hereditary constitution we have not got to take it as an undivided whole, but as something which can be analysed into separate units, just as a chemical substance can be analysed into different elements in definite proportions and arrangement. With this in mind we can turn to another question.

We have discovered in outline how individuals are connected with and related to other individuals of the same species; now we must ask whether whole species are not perhaps connected with or related to other whole species. As Charles Darwin first clearly pointed out, there is always a struggle for existence going on in nature.

From each pair of frogs that breeds, over a thousand fertilized eggs are produced every season; and yet the number of frogs does not increase from year to year. There is, in fact, in every organism an over-production of young, and consequently competition. Since this takes place in all species, there is what we may metaphorically call a pressure of life; the demands upon space and subsistence are greater than the supply, and in every generation only a few of the young produced can reach maturity. There is further the fact of the hereditary transmission of characters, and the origin of new characters by mutation and recombination.

In other words, organisms are continually varying, and some at least of the new variations can be inherited. As an inevitable consequence, if a new inheritable variation appears which is of any advantage whatever in the struggle it will help its possessor to survive, and so will on the average be transmitted to later generations more often than other variations which are useless or in any way harmful.

The new variation may be directly useful in competition with the other members of the species, as for instance if it made its possessor more easily able to secure its prey; or it may remove its possessor from competition with the others of the same species by fitting it to another set of conditions.

In any case, the struggle or pressure of life on the one hand, and on the other the fact that variation does occur and can be transmitted, will lead to what Darwin called *Natural Selection*-the survival of types possessing useful variations, in preference to any other types. A has been stained and is photographed by transmitted light. In it about sixteen muscle-segments have been formed, the three main divisions of the brain are visible, with the eye-vesicles growing out from the fore-brain. The heart can be seen, together with a network of small blood-vessels over the yolk in the outer region of the blastoderm. B has not been stained and is photographed as an opaque object by reflected light.

The amnion has been removed (small traces of it are left in the head-region). The embryo rests on its left side, and the head has bent over. The limb-buds, gill-slits, and allantois can be seen. The blastoderm in one place has rolled over, showing the way the blood-vessels run on its lower surface, next to the yolk. This will explain the fact that animals are adapted to their surroundings, since the unadapted or less well adapted could not survive; it will also explain the fact that when we find a number of species of animals all possessing the same general plan of structure, the different species are generally adapted to live in different conditions, since any variation enabling its possessor to live in new conditions will remove it from competition with the other members of the species.

For instance, the general plan of all frogs and toads is very similar; but some live near water, others in drier places, some altogether in water, others on trees, and so on. The only explanation, in fact, of the resemblance in structure between different frogs and toads is that they are all in very truth related, descended through millions of generations from a common original ancestor; and the explanation of their differences is that they are due to variation and natural selection, the random raw material of the former being sifted by the latter.

The backbone of the evolution theory is that all species have descended from other pre-existing species, and the backbone of the generally current explanation of evolution is that adaptation is due to variation (mutation) acted upon by natural selection. Now frogs and toads resemble newts and salamanders in many points of their organization; for instance, in their moist, scaleless skin, in the structure of their heart, and in their development through a tadpole stage; and accordingly, they are all classed together as Amphibians.

Snakes, lizards, tortoises, and crocodiles, on the other hand, have scales, their heart is a trifle more complicated, and they all develop inside a large-yolked egg with special protecting membranes round them; and they are accordingly all classed as Reptiles. When we classify a number of animals together in this way, we are not merely pigeon-holing them for convenience; we pigeon-hole them in this one particular fashion because we believe them all to be descended from a common original ancestor.

The classification aimed at by Zoology is a Natural Classification, aiming at a grouping of animals according to their blood-relationships. Thus, we believe that all frogs, toads, newts, and salamanders are descended from a common ancestral form with an Amphibian organization; and all snakes, lizards, and other reptiles from a common ancestor with a Reptilian organization. But the organization of Amphibia has many points in common with Reptiles, as also with Fish, and with Birds and with Mammals, including, as we previously pointed out, man himself.

All of these animals possess gill-slits and a notochord at some stage of their development, a backbone, red blood, a nervous system lying along the back, and so forth; they are therefore all classed as Vertebrates (or Chordates, with reference to the notochord), and there is no escape from the conclusion that, in the course of an exceedingly long space of time, to be reckoned in tens or perhaps hundreds of millions of years, they have all descended from an original common ancestor. Belief in the occurrence of evolution, quite apart from any theory of how it has happened, is forced upon us by numerous sets of facts, which remain quite unexplained on any other supposition.

In the first place, there is the fact that animals can be arranged in groups (such as the Vertebrate group), each group with a single general plan of organization; and that within each group there are subgroups with their own special modification of the general plan (e. g. the Birds or the Amphibians within the Vertebrates). Secondly, we have the fact that animals during development often pass through stages in which they resemble other and usually simpler animals of the same group. The frog in the tadpole stage has gills, and even the embryos of fowls and men possess gill-slits. Both frog and man, although tailless when adult, possess tails when young.

All this is very difficult, if not impossible, to explain unless we suppose that our remote ancestors, and those of birds and frogs as well, lived in water and possessed gill-slits and gills. Thirdly, we have the existence of vestiges. Vestigial, or as they are often incorrectly called, rudimentary organs, are organs which are of little or no use to their possessor, although in related forms they are larger and obviously have a function. In man, for example, the appendix, the hairs on the body, the fold of skin at the inner corner of the eye, and the skeleton of the tail are wholly or partly vestigial.

They exist to-day chiefly because our ancestors possessed them in more developed form. The fold of skin in the inner corner of our eyes, for instance, is the remains of the large movable third eyelid of many lower animals. In whales the whole hind-limb has become vestigial. If we look at the hairs on the outer side of our arms, we will find that on the upper arm they point obliquely downwards, on the fore-arm obliquely upwards—all of them, in fact, towards the elbow. In members of the three classes Reptiles, Mammals, and Birds, efficient flying-organs have independently evolved (convergence).

The fore-limb is always utilized as the main part of the wing, and its general plan is retained throughout (homology). But the details are different in each case. The main support (apart from the upper and lower arm-bones) is, in the Pterodactyl, the 5th or 'little' finger; in the Bat, the 2nd to 5th fingers; in the Bird, the quills of the feathers. Accordingly, only in the Bird is the hind-limb not required as part of the support of the wing, and is left free for other functions. That is no good to us. But it appears that some of the apes have the habit of keeping their bodies dry in the rain by clasping their hands in front of their neck. The rain runs down the hairs towards the elbows; these are a little out from the sides, and the long hairs stick out from them to form a kind of spout, from whose end the rain pours off right away from the body.

Fourthly, we have the fact that in species of the same group which are adapted to different methods of life, the same plan recurs in organs adapted to the most different uses. For instance, in the skeleton of the fore-limb in all land animals, although bones may fuse together in some, or altogether disappear in others, the same general plan recurs throughout. This one plan can be seen in the supporting limb of a frog, the paddle-like fore-limb of a whale, the running limb of a horse, the flying limbs of birds and bats, the arm of man. Great difference in detail of adaptation, together with great similarity in general plan—it is difficult to account for this except by common ancestry and a common plan gradually modified in the course of evolution.

Finally, and most conclusively, the fossil remains of animals from earlier periods of the earth's history often show us actual intermediate stages in evolution. For example, the horse has in its forefoot but one finger which it uses, together with two tiny vestigial fingers, the splint-bones. If we accept the idea of evolution, we must suppose that these vestigial fingers were once used; and as a matter of fact, if we look at fossils from a certain period we find no true horses, but animals which, though very like horses in most ways, possess three well-formed toes on the foot.

Again, to take an illustration on a large scale as well as one on a small, if we go back steadily in the history of the earth, we come to a time when man did not exist, or at least no traces whatever of his existence are to be found preserved in the earth's crust. Then, long previously, to one when all the mammals were small and primitive, and all the birds toothed; then, to a time when there were no mammals or birds at all, but great reptiles, many of types now unknown, were the dominant living things; before that, to a time without reptiles, when amphibia were the only land vertebrates; and before that again, to a world without land vertebrates at all, but still with fish in the sea; and finally, a stage is found in which there are no fish, but only marine invertebrates.

The study of heredity showed us that each individual arose from actual portions of living substance which had once formed part of other individuals.

The study of evolution shows that species arise from other species. Within a single species there are a number of parallel streams of living substance flowing through the generations; but these parallel streams may diverge, and the original species branch into two. Since each species has evolved out of another species, and each individual grows from a detached part of another individual, the whole of life must be looked on as a single mass of living substance, flowing on in time, and divided in the course of its history into a number of separate courses of different size, the main groups, the smaller groups, and the species. The alterations in the branches of this mass of living substance we call Evolution.

11

Disease Detection and Elimination

In micropropagation, the health status of the donor mother plant and of the plants multiplied from it are among the most critical factors, which determine the success of a tissue culture operation. The indexing of the mother plants for freedom from viral, bacterial, and fungal diseases is a normal procedure in large-scale plant propagation through tissue culture. Plants not originating from pathogen-tested material must be screened for the presence of viruses Laboratories, which do not have in-house facilities to carry out plant indexing, should obtain their indexed stock plants from organizations such as Departments of Agriculture, agricultural universities or privately owned certified germplasm repositories that routinely produce such plant material.

Batches of micropropagated plants should be tested for freedom from diseases either in-house or by other laboratories. ELISA has been the most effective method for virus and pathogen detection in plants. Polymerase chain reaction (PCR) and nucleic acid hybridization are more sensitive than ELISA, and can detect pathogens in extremely low amounts. The elimination of most viruses can be achieved by a combination of apical meristem culture and thermotherapy.

The health status of the donor mother plant and of the plants multiplied from it are among the most critical factors, which determine the success of a tissue culture operation. Hence, indexing of the mother plants for freedom from viral, bacterial, and fungal diseases is a normal procedure before undertaking propagation in large-scale plant propagation through tissue culture. Laboratories, which do not have in-house facilities to carry out plant indexing, should obtain their indexed stock plants from organizations such as Departments of Agriculture, Agricultural Universities or privately owned certified germplasm repositories that routinely produce such plant material. Batches of micropropagated plants should be tested for freedom from diseases either in-house or by other laboratories.

Bacterial and Fungal Contamination

While most of the fungal and bacterial diseases are eliminated during surface sterilization and culture, viruses and viroids survive through successive multiplication if the mother plant is infected. Even cultures from seeds may carry viroids and bacterial endophytes, in certain cases also viruses. Bacteria may be controlled under in vitro conditions to some extent using certain bacteriostatic agents. Nevertheless, such chemical treatments with *e.g.* 'plant protection mixture' (ppm) or commercial antibiotics even when applied on long term do not fully eliminate microorganisms.

Virus Infection in Plants

Viruses are obligate parasites, and for their multiplication, they use the metabolic machinery of the cells, and inhibit plant growth. Although, many viruses do not produce disease symptoms, they adversely affect plant metabolism and increase progressively with repeated vegetative propagation. Most viruses infect only a limited number of species, but a few have a wide host range.

Many plant viruses are transmitted through vectors such as insects, nematodes, bacteria, and fungi. Propagation from vegetative parts, grafting and sap can transmit viruses from one plant to another. The damage from viruses depends on their concentration in the tissues and spread within the plants.

The greater the concentration of the virus, the greater the debilitating effect observed on plant growth. Many varieties of vegetatively propagated crops decline in performance with viral accumulation and must be discarded. Even though the viruses do not produce visible symptoms, the plants have reduced yield and poor quality. The removal of specific viruses by meristem-tip culture has been reported to lead to a dramatic yield increase and rejuvenation of plant varieties that are vegetatively propagated. This is a unique contribution of meristem-culture not achievable by any other techniques.

Detection of Viral Diseases

Plants not originating from pathogen-tested material must be screened for the presence of viruses. Procedures to detect the presence of viruses include visual examination for viral symptoms, infection tests on indicator plants, serological tests, electron microscopy, and direct detection of RNA using molecular techniques. Internationally approved and recognized detection systems include serological and molecular laboratory assays, and indicator hosts in the greenhouse.

Indexing with Indicator Plants

Prior to the advent of antigen-antibody reaction detection systems, the earlier method of detecting virus relied upon indicator plants. For example, in potato, viruses A, X, and Y can be detected by rubbing the juice of the infected plant on the leaves of indicator plants such as Chenopodium spp., and Nicotiana spp. The development of necrotic spots on the leaves is a typical symptom indicating the presence of the viruses. For example in sweet potato, its wild relative, Brazilian Morning Glory (Ipomaea setosa) is still used as the indicator host. Similarly, in many fruit trees, the appropriate rootstocks are used for indexing of various virus strains. The detection of plant diseases has changed radically in the last 20 years. A number of indicator plant based tests have been replaced with specific Enzyme Linked Immuno Sorbent Assays (ELISA). Nevertheless, indexing with indicator plants is a low cost and highly effective method to detect many viruses.

ELISA Test

ELISA has been the most effective method for virus and pathogen detection in plants. Each virus has a unique protein coat. Hence they can be detected by using antibodies, which are usually linked to enzymes for reporting the affinity binding via colour signal from the used enzyme substrate. The double antibody sandwich (DAS) method in a 96-well microtitre plate is still widely used. Various alternative formats, substrates, and antibody binding modifications have been developed over the years to increase specific sensitivity. The most common method nowadays uses "biotinylated" virus-specific antibodies to improve their affinity to the respective virus surface thus improving the signal. The use of monoclonal antibodies has been so far restricted to certain potato and barley viruses, because of their high cost and inability to detect the range of strains in several diseases.

Large scale testing based on screen-spot samples is also a routine procedure. Such dotimmuno- binding assays are versatile and cheaper alternatives to microtitre plate tests. Plant sap is directly spotted onto a pre-treated microporous membrane (*e.g.* nitro-cellulose or polyvinylidene difluoride). The test kits can be used directly in the field. Sample membranes can be dried and shipped to another laboratory for processing. However, such systems are not useful for viruses present in low concentrations in the plant sap.

ELISA has several limitations. First of all, a test is specific only to a given strain of the pathogen. For a large group of ubiquitous viruses, which do not produce disease symptoms or fatal consequences, ELISA kits have not been developed. Thus, the strain specificity has severe limitations for comprehensive quality control. Further, ELISA is less sensitive than PCR, and may fail to detect low amounts the virus in tissue-cultured plants. Whereas a positive

ELISA is a good indicator of the existence of microbes (but false positives may also occur), a negative test does not ensure their absence. The use of Immuno-Tissue-Printing allows the localization of viruses in tissues and therefore the improvement of elimination strategies in vitro by visualizing the result of virus removal.

Polymerase Chain Reaction

The Polymerase Chain Reaction (PCR) is more sensitive than ELISA, and can detect pathogens in extremely low amounts. PCR detects pathogens from their genetic material, *i.e.* DNA or RNA. This technique uses a specific enzyme and the respective genomic start code for a relevant section of the pathogen-DNA to reproduce millions of copies from it. Therefore, even minute quantities of the pathogen DNA can be detected. Since most plant viruses are made of only RNA, first a DNA copy of the RNA is made, whereas DNA from bacteria and phytoplasmas may be amplified directly. Because the genetic material of the pathogen is mixed with that of the plant, it is necessary to target only the genetic material to be copied. This is done by adding small pieces of DNA called primers which stick only to borders of the region of genetic material that need to be copied. The primers then direct the enzyme where to start and end amplification.

During PCR, the DNA is subjected to a series of hot, cold, and warm cycles in a thermocycler. Hot cycles split DNA into single strands. Cold cycles allow the primers to attach to the DNA. During warm cycles, the enzyme makes a copy of each piece of primed DNA. If a phytoplasma is present, PCR produces a large quantity of a specific piece of DNA. By using gel electrophoresis, DNA pieces are separated in size, and the specific piece of DNA is identified.

If the critical concentration of a DNA particle is sufficiently low and the sample tests negative, the plant material can be regarded as practically free of the target microbe. Once a test is standardized with an appropriate primer, and by a method, which excludes inhibitors, a negative PCR test is a more reliable indicator of practical freeness. The use of PCR drastically reduces the number of samples, which test negative although they may carry the virus. Once proper probes are developed, the nylon membrane spot test is a practical, rapid and accurate method for diagnosis. PCR has the advantage that by determining "degenerate primers", a fairly large group of microbes such as viruses, bacteria and fungi can be detected.

Nucleic Acid Hybridization Test

The presence of viruses can be detected from isolation and characterization of double stranded RNA (ds-RNA) produced in most plants during viral replication. Although nonspecific presence of ds-RNA in a sample extract strongly suggests viral infection, for further identification of the virus, the test has to be verified by serological and other methods. Viroids are much smaller than viruses, and are made of a circular piece of RNA and do not have a coat protein. Viroids such as 'Potato Spindle Tuber Viroid' can be detected by nucleic acid hybridization. The RNA of the virus "hybridizes" with probe RNA. When the two single strands come together, they form a double strand.

To detect the viroid, RNA extracted from plant tissue is attached to a microporous membrane. This bound RNA is then detected by a "probe", which is an RNA particle labeled with either 32P or a chemical called DIG (digoxigenin). If a viroid is present, the probe hybridizes with it; if not, the probe is lost when the membrane is washed. DIG bioluminescence or radioactivity if present on the membrane can be detected on an exposed film. Samples from known healthy and known infected plants are used as controls. Large scale testing on nylon screen-based spot samples is also a routine procedure.

Elimination of Viruses

The elimination of viruses can be achieved by a combination of apical meristem culture and thermotherapy. Meristem culture is the most commonly used method to free plants from virus diseases. This should not be confused with shoot tip culture as explained before. Virus-free plants have been obtained by thermo-therapy and meristem culture in several species. The plant meristem is a zone of cells with intense divisions, situated in the growing tip of stems and roots. The virus travels through the plant vascular system, which is absent in the meristem. Moreover the cell-to-cell movement of the viruses through plasmodesmata cannot keep up with the growth and elongation of the apical-tip. The high metabolic activity of meristematic cells, usually accompanied by an elevated endogenous auxin content in shoot apices may also inhibit virus replication.

Thus, the meristem is highly protected from infection. Based on this finding, meristem culture has been extensively used to eliminate viruses, bacteria and fungi from plants. The culture of meristems or alternatively small shoot tips, in combination with enhanced cell division in vitro and/or thermal pre-treatment allows the elimination of viruses in plants propagated from vegetative parts. If explants are too big, they are likely to contain virus particles in the associated vascular tissue.

For thermotherapy, the plants are first grown at high temperature (ca. 38-42° C) for 4-6 weeks. Under tropical or subtropical conditions, this can be accomplished simply by installing a small compartment of a glasshouse equipped with a roof vent on one end and an exhaust fan on the opposite end, both temperature-regulated. This approach removes the excess heat and provides a constant high temperature daytime treatment. In climates with temperate conditions, the same effect can be achieved by placing fluorescent lights including ballasts and/or heat producing incandescent lamps in the necessary minimum distance to avoid damage over the plants to be treated in a dark box just large enough to include the plants. Such a system has been used for virus elimination in sweet potato.

After thermotherapy, 0.2 –0.4 mm explants are preferentially cultured singly in test tubes. If the explant is too big, it likely has a vascular system that may contain microbial contaminants including viruses. The plants thus obtained are multiplied, and re-indexed. A better strategy is to culture 2-5mm long explants for 4-5 weeks, maintain the in vitro grown plant at high temperature for 4-5 weeks, and excise 0.2-0.4mm or even longer explants to initiate subcultures.

This procedure avoids in vivo contamination problems and gives high survival and multiplication rate. In potato, in vitro cultures were established from several millimeter long shoot tips and axillary buds after in vitro thermotherapy to free potato from all relevant virus diseases. By using this method, it was possible to eradicate viruses A, Y and X from potato varieties in one single step.

A method called multiple lateral shoot technique of in vitro elimination of three common viruses, X, Y and S of potato, has been also reported. In this method, a stem with at least five nodes is treated with ribavirin in liquid medium, and cultured for 5 days at room temperature. The stems are subjected to thermotherapy at 32-35° C for 25 days, apical buds are taken from the lateral shoots cultured on solid medium, and checked for virus infection with ELISA. The survival of explants in relation to the applied temperature stress and their size is always inversely proportional to the pathogen elimination success. It is known from molecular and traditional microbiological surveys that effectively all in vitro plants contain certain microorganisms, mostly bacteria but also mycoplasmas, viroids, and fungi, many of which cannot be cultured without the host. Such microorganisms may not show symptoms, and in some cases may even have a positive effect on the growth performance of the host plant.

Shoot-tip culture is used for the multiplication of plants that are already freed from known diseases. It should be emphasized that it does not free the plant from viruses. In fact, it favours propagation of viruses and increases virus concentration in the daughter plants. In many ornamentals, variegation is due to the presence of certain viruses or mycoplasms. The removal of such viruses is therefore not desirable. In such ornamental plants, multiplication from shoot-tip and axillary bud culture is the better choice, because meristem culture may remove the virus, and the ornamental variegation is lost. In certain varieties with variegations as in Petunia and Pelargonium meristem culture may dissolve the existing chimera, thus producing material without the desired character.

Viruses may also produce certain phenotypes that are characteristic of a plant variety. For example, in sugarcane cv. 'Co 740', the yellow irregular leaf streaks were initially considered as a diagnostic character till they were proven to be disease symptoms. The removal of the virus by meristem-tip culture led to the disappearance of the yellow streaks, which reappeared on re-infection. Erroneously, the yellow streaks are still recognized as a varietal diagnostic trait of 'Co 740'. Recently, endogenous para-retroviruses have been reported that integrate in the genome of the plant material. Under in vitro stress, sequences of such viruses have been detected in the micropropagated banana plants.

Perspective of Disease Freeness

Despite the fact that many viruses may not show visible symptoms, their presence can reduce the yield and quality of crops. Yield increases of up to 300 per cent were reported following replacement of virus-infected stock with specific pathogen-free plants. In China, yield losses from virus diseases of sweet potato (Feathery Mottle Virus, SPFMV, Latent Virus, SPLV and Caulimo-Like Virus, SPCLV) used to be as much as 20 per cent. After the establishment of a virus elimination programme, within a few years almost 10 per cent area of 6.6 million ha under sweet potato is now planted with virus-free planting material.

There are no chemicals to cure virus-infected plants in the field. However, some less virulent viruses can protect plants from infection of their more virulent strains and create cross protection. The phenomenon of cross protection need not prevent the use of meristem culture. Under field conditions a crop does not stay absolutely virus free even though derived from tissue culture material. Even under tissue culture, total freeness may not be achieved. Many symbiotic organisms contribute to the vigour and ability of the tissue-cultured plants to perform well in the field. In the course of tissue culture, many plants are made free from such beneficial microbes, *e.g.* symbiotic nitrogen fixing endophytic bacteria, and mycorrhizae. There is no practical way to retain the beneficial microorganisms during tissue culture. The deliberate re-infection of propagules with selected strains can be a valid way of retrieving the benefits of such microbes.

Quality Assurance

Quality checks are essential to assure production of high quality plants and to have endusers confidence. Quality standards require the establishment of suitable tests to maintain quality control. The choice of explant source, freedom of the donor plant from viruses, disease causing fungi, bacteria, viroids, phytoplasmas, vigour and conformity of the variety, and elimination of somaclonal variants are critical for maintaining plant quality. Variety identification by proper labeling at all stages is essential to ensure varietal identity.

In many developing countries, there is a major need to provide large quantities of improved planting material to growers and farmers. Since quality checks in many developing countries do not exist beyond the laboratory gate, the assurance of high quality micropropagated plants is critical to have end-users confidence. Quality assurance becomes even more important when the plant material is for export. High quality of the super-elite and elite micro- and mini-tubers of potato has improved the production of quality seed potatoes in Ireland and Scotland where conventional multiplication had been inadequate to generate high quality disease-free planting material.

Cost reduction based on low cost options is necessary for commercial micropropagation. While cost reduction is necessary for commercial micropropagation, the low cost option should not compromise quality. The sturdiness of plants is an important attribute of quality, which determines their ability to adapt to field conditions. The cost-effective systems should produce semi-hardened plants in the culture itself. Such plants produce a profuse and viable root system and adapt in less time to field conditions.

Quality standards require the establishment of suitable tests to maintain quality control and the assurance of micropropagated material irrespective of the system used. No industry can prosper without a quality policy based on carefully identified, relevant and recognized quality specifications and standard methods to test and verify them. Hence, the plants coming out of any facility should be subjected to quality testing without discrimination or prejudice. The commercial micropropagation industry, which is still in its infancy and early stages of evolution in many developing countries, lacks this vitally important component. Under the seed law, tissue culture plants are defined as "seed".

The specifications stipulated for seeds of cultivated crop plants do not apply to tissue culture plants for quality, except for the basic requirement of distinctiveness of the variety, genetic uniformity, and stability. This lack of fundamental specification creates conflicting opinions on the quality of micropropagated plants. Such a situation is undesirable for users, producers and regulating agencies and a hindrance to the maturity and growth of commercial micropropagation as a vibrant industry. Several commercial micropropagation laboratories already follow standard practices such as labeling of cultures, use of indexed stock mother plants, and preventing secondary infection from pathogens during the hardening phase.

The quality considerations need to take into account several parameters as a package. A comprehensive approach is needed, which will help micro-propagators to monitor production quality, and formulate a quality- assurance system based on such tests and specifications, that are open to third party inspection. Such an approach will also be useful to formulate well-defined quality specifications by regulatory authorities.

Factors Affecting Quality

The explant source (*e.g.* meristem tip, axillary buds), freedom of the donor plant from viruses, disease causing fungi, bacteria, viroids, phytoplasmas, vigour and conformity of the variety, and elimination of somaclonal variants are critical for maintaining plant quality. Many cultures are initiated from meristem-tip explants. Such cultures can be multiplied by enhancing axillary branching, which preserve genetic stability even on indefinite subculturing. Since this method reduces the chance of differentiation of plants from cells other than the meristematic tissue, the expected genetic variation is low.

However, the shoot-tip and bigger explant cultures have differentiated cells, in addition to the undifferentiated meristematic cells, and may result in somaclonal variants. Thus, shoot tip culture has a significant probability of producing genetic variants. However, in banana, the frequency of such variants markedly differs with the variety. Shoot-tip explants infected with endophytic microflora and viruses transmit them to daughter plants. Endophytic microflora is not desirable, unless it has symbiotic functions. Hence, meristem culture is the preferred method to initiate cultures for producing high quality plants.

Multiplication should be based on enhanced axillary bud proliferation, *i.e.* the rapid formation of tiny microscopic axillary buds in the axils of the leaves of the tiny plants that sprout and develop into daughter plants. The introduction of callus phase usually stimulated with growth regulators such as 2,4-D will introduce unwanted somaclonal variation and should be avoided during multiplication. For the same reason, the use of relatively high concentrations of growth regulators such as BAP, IAA, and BA should be avoided.

Quality assurance begins with culture initiation, explant source, and judicious use of growth regulators during multiplication. Cultures initiated from meristem tip and multiplied through enhanced axillary branching avoid callus formation and give uniform clones. On the other hand shoot-tip cultures are considered inferior to meristem-tip culture since they can be the source of virus infection. In case of photoperiod-sensitive plants, light duration can also have a serious effect on the final performance of the micropropagated plants, particularly those meant for flower production. Hence, a neutral photoperiod of 12/12hr light/dark is preferred during multiplication.

Freedom from Diseases

It is necessary to initiate cultures from the meristem-tip and ensure further multiplication free from viruses. For example, the badna virus in banana integrates into the host genome during tissue culture and can become a source of infection in micropropagated plants although parents are free from symptoms. If the parent plant is virus infected, the cultures from its shoot-tip will likely be infected. Even the meristem tip will carry over viruses unless thermo-therapy has been carried out and the cultures have been tested virus free. Removal of viruses will be positively undesirable in exceptional cases of ornamentals, where the characteristics are associated with the presence of viruses, as in the case of the yellow vein geranium.

The quality tests should be rapid, and aim at the identification of a variety or clones, and ignore point and bud-mutations, which occur in the normal population as a natural phenomenon. A variety is an assemblage of closely similar genotypes and clones, and not a group of "absolutely uniform" genotypes. Thus, the methods should be able to make a distinction between two varieties, and not produce confusion with regard to intra-varietal variability. The tests for quality include endophytic bacterial count. Broad range identification methods for the presence and quantification of endophytic bacteria are required and should be the first tests in quality control. This will prevent production of plants contaminated with undesired endophytic bacteria, and may indirectly be a useful indicator of the absence or of low concentration of viruses.

Conventional bacteriological methods are enough to standardize such a test. PCR based methods may provide identification of a wide range of bacteria. For the identification of viruses, light microscopy for inclusion bodies has limited sensitivity. Immunology based methods have been commonly used for detecting specific viruses *e.g.* in potato. However, PCR methods are being increasingly developed and finding increasing use.

Labeling and Culture Identity

Micropropagated varieties in culture and at the hardening stage are difficult to distinguish visually from one another. Variety identification by proper labeling at all stages is essential to ensure varietal identity. When several varieties of the same plant species are used, accidental mislabeling or mixing may occur. Therefore, monitoring is essential to detect admixtures at all stages. Clones of the mother variety should be maintained by conventional vegetative propagation to fall back upon whenever necessary.

Varietal Conformity

Varietal conformity based on morphological characters has been the basis for identification in conventional propagation and the same applies to the plants derived from tissue culture. However, considerable caution is advised in selecting morphological characters for varietal conformity. For example, in sugarcane, variation in morphological characters such as cane colour, shape of the bud, size of the ligule can vary in the different parts of the same cane. Where morphological characters of mother plants of diagnostic importance are due to interaction with microbes, meristem tip culture invariably gives rise to a new phenotype due to removal of such microbes. In such cases, the new virus free phenotype represents the original expression of the new virus free phenotype. Hence, criteria other than morphology are necessary for varietal conformity. Morphological characters are not an absolute proof of genetic conformity or genetic stability in tissue culture. Many times, physiological changes may lead to the appearance of types among micropropagated plants that differ from the parental clone. In some cases, such variants with minor phenotypic changes can be used for field production, but not for propagation. Low frequency mutations and bud mutations occur in conventional populations propagated from vegetative parts. Genetic changes not associated with morphological traits can be detected by DNA based molecular methods. Somaclonal variants are undesirable in cloning, but can be used for breeding.

DNA-Based Molecular Markers

DNA-based molecular markers are not affected by the environment and are more reliable than morphological characters for identification of varieties but are not a low cost option for small facilities. Hybridization and RAPD (Randomly Amplified Polymorphic DNA)-based methods can discriminate between varieties and clones. However, the differences between the donor plant and those regenerated through tissue culture cannot be detected with such methods. Only in a few cases, it is possible to discriminate the mother plants from the variant regenerated plants. Molecular methods appear to be reliable for discrimination at the variety level. Variety identification based on DNA fingerprinting, such as RAPD, microsatellites, STR's (Sequence Tagged Repeats), AFLP (Amplified Fragment Length Polymorphism) and multi-locus probes, is potentially useful. RAPD can be standardized within a short time without elaborate knowledge of DNA sequence. However, limitations in the reproducibility of this method (between two sets of experiment, two machines, two workers and two laboratories) are well recognized. This limitation can be overcome by including a sample of reference DNA. Epidermal patterns and structures and stomata, if stable in tissue culture, are additional tools in variety diagnosis. If their validity is confirmed, they may be simpler than RAPD and other molecular methods.

Cost and Quality

For small facilities, molecular tests are expensive. For them, it is cheaper to obtain the services of large or centralized laboratories (*e.g.* departments of agriculture and universities), which can either index cultures for a fee or supply indexed mother cultures at regular intervals. Unequivocal tests can help to develop in-house quality control systems. Such tests can be used for developing third party inspection systems, and provide a practical basis for statutory quality control systems.

Priming Tissue Cultured Propagules

Priming of in vitro propagules refers to the manipulation of the growing environment, prior to and upon transplanting, and is an integral part of tissue culture propagation. Tissue-cultured propagules are produced under controlled environment. Such plantlets have small juvenile leaves with reduced photosynthetic capacity, and malfunctioning stomata. Priming for rooting, shooting, and improved photosynthesis can be achieved with growth regulators and simple adjustment to the growing conditions that affect the post-transplanting performance of the propagules.

Vented closures with microbial filters have been used to facilitate gas exchange to reduce ethylene build-up that stunts plant growth, reduces leaf-size, and causes leaf drop in tissue culture containers. Plantlets produced under photoautotrophic culture systems on media with or without sucrose but CO_2-enrichment, increased light intensity, good gas exchange and reduced humidity are more vigourous, have larger root- systems, and are less susceptible to microbial contamination. Plants adapted gradually to the ex-vitro environment have improved survival upon transfer to soil. Plants in their natural environment are colonized with many bacteria, fungi, and mycorrhizae. In-vitro or ex-vitro biopriming of micropropagated plants with such organisms can improve plant performance under stress environments, and consequently enhances yield.

The adaptation of plants to the environment-temperature, osmotic stress, and pathogens, is triggered by modification of expression of specific genes and metabolic pathways. Certain chemicals, environmental factors, and micro-organisms pre-sensitise the cellular metabolism of plants. Upon exposure to stress, the pre-sensitised or primed plants adapt better and faster than the non-primed plants. The production of high quality and vigourous plants through in vitro culture requires the enhancement of post-transplanting ability for water management, efficiency of photosynthesis and resistance to diseases. Priming of in vitro propagules, based on manipulation of the growing environment (chemical, physical and biological) prior to and upon transplanting, is an integral part of tissue culture propagation. The ability of the propagules to withstand transplanting stress very often determines the success or failure of tissue culture operations.

Tissue-cultured propagules are produced in controlled environments under pre-set temperature and low-light intensity, and in small vessels with poor air exchange. This leads to high humidity and accumulation of ethylene. Also, the nutrients in the growing media, sucrose, mineral salts, vitamins and growth regulators, are far in excess of those in the soil. Such an environment causes developmental distortions, and represses or modulates several metabolic pathways. As a result, the plantlets have small juvenile leaves with reduced photosynthetic capacity, and malfunctioning stomata. The roots of such plants lack hair or have few, and the leaves have poor cuticle development and low wax deposit.

Priming for Rooting

The ability to manage water, photosynthesis and response to stress during and after in vitro culture determines the final performance of the propagules. The formation of a functional root system is an essential step to manage water under the post vitro conditions. The in vitro roots remain functional and continue to grow during ex-vitro acclimatisation. Easy to root species, (*e.g.* potato) usually regenerate roots upon subculture on hormone-free medium. However, root architecture and biomass partitioning between roots and shoots can be easily manipulated by simple adjustment to the growing conditions that affect the post-transplanting performance of the propagules. For example, increasing day/night temperature from 20/15 to 33/25°C either slightly stimulates or has no effect on the shoot growth in clones of potato, but dramatically enhances root growth and root branching.

Several chemicals also influence the development of the root system. Abscisic acid is known to affect initiation of new lateral roots and overall changes to the root architecture Jasmonic acid (JA) has a similar effect. Low JA concentrations (0.5- 1.0 μ M) stimulate both root and shoot growth. Jasmonates induce stress resistance in plants, and may be useful for priming of in vitro plants during the last subculture.

In potato, nodal explants from JA primed stock plants tuberize earlier and more uniformly, and produce more microtubers than the non-primed controls. Salicylic acid (SA) is a major signalling compound and induces abiotic and biotic stress resistance in plants. It also induces adventitious roots in cuttings and stomata closure, and inhibits ethylene biosynthesis in detached leaves; hence it may also be useful for priming tissue-cultured propagules. Many woody and herbaceous perennials require auxins for rooting. Auxins are involved in root initiation and adventitious root formation. However, excessive auxins after root induction may inhibit root regeneration and elongation, and impair formation of a good root system. Therefore, a careful selection of the auxin type and concentration is recommended. IAA and IBA are usually used for easier-to-root plants and NAA for more recalcitrant woody plants. The efficacy of different auxins also depends on the explant type, and exposure to light. For example, a short duration (3-7 days) of dark incubation enhanced rooting in apple, and papaya.

Addition of low concentration of the growth retardant daminozide to hormone-free medium enhanced post-transplanting acclimatisation of potato plants. Other supplements like activated charcoal, seaweed concentrate, pyroligneus acid, and dilution of salts in the basic MS medium can also aid root initiation and growth. In vitro cultured papaya shoots in aerated vermiculite gave better rooting with higher post-transplanting survival than on non-aerated vermiculite and non-aerated agar. In some difficult-to-root species, *e.g.* apple and Saskatoon berries, in vitro rejuvenation via several subcultures is required before successful rooting of microcuttings. Individual cultivars may require adjustment to culture conditions and different number of subcultures.

For example, over 90 per cent rooting was achieved with apple microshoots, cv. 'Jonathan' by the ninth subculture, but a more recalcitrant variety 'Red Delicious' required over 30 subcultures for successful rooting. In Saskatoon berries, where rooting by traditional stem cuttings is very difficult, tissue culture rejuvenation and subsequent transplanting to the greenhouse can generate highly vigourous plants. Shoot cuttings taken from such planted hedges readily root within 2-3 weeks after a simple treatment with IBA.

Increasing Photosynthetic Competence

Tissue culture containers provide sterile conditions and high moisture, thereby preventing contamination and desiccation of the explant, but restrict aeration and escape of metabolically generated ethylene and water vapours. Such conditions slow down growth and induce undesirable morphological changes. The ethylene build-up stunts plant growth and reduces leaf-size, and causes leaf drop. Such plants desiccate quickly upon transfer to the external environment. Vented closures, including those for baby-food-jars with built-in different sizes of microbial filters, have been used to facilitate gas exchange between culture vessels and the ambient environment.

High relative humidity in culture vessel reduces deposition of epicuticular waxes on leaves. It impairs the functioning of stomata and alters leaf mesophyll by reducing the number of palisade cells and by forming spongy parenchyma with large air spaces. Other effects include low nutrient uptake due to decreased transpiration, decreased lignification, and reduced or abnormal development of the vascular system. Although the cuticular wax development and stomata functioning can be induced by gradual adjustment to lower humidity, morphological changes cannot be easily reversed. The altered leaf morphology reduces photosynthesis and enhances respiration. Cooling the bottom of culture vessels by 2-3°C below the ambient air temperature in the culture room or providing high light intensity can reduce high humidity. Gradual opening of vessels for a few days prior to transplanting can also adjust high humidity. Many micropropagated plants have C3 photosynthetic pathway; hence, are greatly affected in their ability to assimilate carbon. The balance between photosynthesis and photorespiration controls their productivity. Under the traditional methods of micropropagation, cultures are usually mixotrophic (use carbon source mostly from sucrose in the medium as well as photosynthesis). The carbohydrates in the media inhibit carbon assimilation via photosynthesis by reducing rubisco (ribulose-1-5-bisphosphate carboxylase/oxygenase) activity. On the other hand, several studies indicate a stimulatory effect of sugar on biomass accumulation under elevated CO_2. In the photoautotrophic culture systems the cultures are grown under CO_2-enrichment (1000-1500 μ mol mol^{-1}), increased light intensity (at least 150 μ mol m^{-2}, s^{-1}), good gas exchange and reduced humidity, and on media with or without or sucrose. Plantlets produced under such conditions are more vigourous, have larger root- systems, and are less susceptible to microbial contamination. They are able to photosynthesize, and do not require acclimatization during post-vitro establishment.

Priming for Temperature Tolerance

In vitro grown plantlets can be acclimatized to low temperatures for cold tolerance under field conditions. Alfalfa and red clover seedlings withstood -16 and -10° C, respectively, upon cold hardening at 2-5° C, 80ìE m^{-2}, s^{-1} light and 8h photoperiod, for 3-5 weeks. Such hardening has been successfully utilized in the selection of cold tolerant alfalfa and red clover genotypes. Elevation of sucrose in the culture medium or application of abscisic acid, or both, may further improve plantlet hardiness in some species.

Acclimatization to Soil Transfer

Good quality propagules with well developed roots and leaves are easy to acclimatize to the external environment. Any successful acclimatization protocol must ensure that the plants maintain active growth during the entire weaning period. The effect of in vitro growth conditions on post-vitro acclimatization has been demonstrated in grapevine and chile ancho pepper, Capsicum annuum. The most common approach to improve plant survival upon transfer to soil is their gradual adaptation to the ex-vitro environment. Under such conditions, plants rapidly convert from heterotrophic or photomixotrophic to autotrophic growth, form fully functional root systems, and their stomatal and cuticular transpiration is reduced. Such gradual adaptation is often carried out in greenhouses by decreasing relative humidity using fog or mist chambers and increasing light intensity and shading techniques.

Light intensity needs to be adjusted to the plant requirements. Rooting of in vitro produced shoots in vivo is recommended for some species to promote development of roots already adapted to external conditions, to prevent potential damage to fragile in vitro roots, or to avoid post-rooting dormancy. Other treatments of in vitro plantlets upon transplanting to the outside environment include high phosphate fertilizer to enhance their vigour, fungicides to prevent damping-off and anti-transpirants to prevent desiccation; however, they may cause phytotoxicity.

A simple, low cost method for ex vitro weaning and liner production of woody plant species and Hosta has been developed. The shoots with root-initials from the rooting medium are transplanted in 'Jiffy' plugs. Plants are acclimatized in plastic trays (25 × 50 × 5cm) with transparent cover. Each tray holds 300 plugs, each 18mm diameter. The plugs are saturated in a 2.5-3L solution of fertilizer (1g/L) 10-52-10 NPK; the excess solution is discarded before transplanting. The plants are watered weekly (approx. 1L water/week) and kept in the growth room benches for 3-weeks after transplanting at 22-24/20°C day/night, 14h photoperiod, under 120 μ $Em^{-2}s^{-1}$ fluorescent light. The current cost is US$73 for 5400 plugs, $58 for 100 trays, and $60 for 100 plastic covers; the total cost per plug being US$0.018. The rooting success in woody plants, *e.g.* Swedish columnar aspen, Saskatoon berry, chokecherry and pincherry cultivars, sour cherry, low bush blueberries and others was more than 95 per cent, and 100 per cent with hosta varieties. In seed potato production, 2 to 4-week old tissue cultured plantlets are usually hardened off for 2-4 weeks in a greenhouse, and transplanted in the field. Such plantlets are still juvenile and not well adapted to stress. To prevent desiccation, insect and bird damage, and spread of viruses by aphid, many growers cover plants with floating covers and install an overhead irrigation system. Seed potatoes produced under floating covers are consistently better in quality than those from the unprotected planting.

In vitro produced-propagules of woody species, *e.g.* Saskatoon berry (Amelanchier alnifolia Nutt.), Chokecherry (Prunus virginiana L.), Pincherry (P. pennsylvanica L.), Swedish columnar aspen (Populus tremula var. erecta), European weeping birch (Betula pendula var. 'Gracilis' become dormant after rooting. In vitro rooted Saskatoon berry is particularly prone to dormancy as stress factors readily induce terminal bud dormancy. When the shoots stop growth, leaf shedding occurs in the culture or soon after transplanting, and the propagules desiccate. The liners produced from such plants are less vigourous and difficult to establish in the field.

In Chokecherry (P. virginiana L.), dormancy symptoms of shoots with reddish leaves and cessation of terminal-bud growth are induced upon depletion of nutrients in the medium. Post-rooting dormancy has also been reported in rhododendrons and low bush blueberries. Adding low amounts of gibberellic acid (GA) to the medium prevents dormancy, although shoots treated with GA are difficult to root. In Saskatoon berry, ex-vitro rooting under non-sterile conditions is helpful, although does not eliminate the problem. Spraying leaves of ex vitro rooted-shoots with BAP (400ppm), followed by GA (100-PPM) the next day prevents dormancy.

Priming Somatic Embryo-Derived Plants

Somatic embryos are produced as adventitious structures directly on explants of zygotic embryos, from callus and suspension cultures. The stress inducing molecules in plants, ABA and JA have been successfully used to facilitate embryo maturation and priming for resistance to post transplanting stress. ABA and osmotic shock with 6 per cent sucrose is a routine step to induce desiccation tolerance in alfalfa synthetic seeds. Osmotic stress induction with 3 per cent sorbitol was effective in embryo maturation and conversion in soybean cultures and ABA treatment combined with partial desiccation of encapsulated somatic embryos in sugarcane.

Several sugar alcohols and polyethylene glycol have been tested with different degrees of success for priming SE propagules of various plants. In conifers, post ABA embryo maturation has been significantly improved by benzyl adenine (BA). The lowering of sucrose concentration from 58 to 29mM in the proliferation medium improved the formation of cotyledonary SE on the maturation medium. Interaction between plant growth regulators and various carbohydrates and organic nitrogen compounds in embryo development and maturation has been reported.

Biopriming

Plants in their natural environment are colonized both by external and internal microorganisms. Some microorganisms, particularly beneficial bacteria and fungi, can improve plant performance under stress environments, and consequently enhance yield. Plants infected by microorganisms develop systemic resistance (systemic acquired resistance, SAR, or induced systemic resistance, ISR), and/or benefit from their antagonistic abilities towards pathogens (cross protection). Although, the inoculation of seeds with beneficial microorganisms has been practiced for more than 50 years, the inoculation of tissue culture propagules to enhance plant performance is relatively new. Plant tissue culture is based on axenic (contaminant-free) culture systems. Hence, microorganisms, including endophytes, are treated as problemcausing contaminants, and various procedures have been developed to eliminate them. Only recently, have microbial inoculants, primarily bacterial and mycorrhizal, been evaluated as propagule priming agents both as in vitro co-cultures and on transplanting.

Standard microbiological techniques allow culturing only a few per cent of the naturally occurring microorganisms. Moreover, the soil and plant associated bacteria can switch between culturable and unculturable stages. The development of new culture methods that allow the establishment of stable associations between plants and beneficial organisms, both in vitro and ex vitro, under different environments is a major challenge of future research.

In Vitro Bacterization

Rhizosphere bacteria interact with plants and other inhabiting organisms by producing a wide array of chemicals. Many of these chemicals exhibit plant regulatory activity. There is increasing evidence that the bacteria associated with plants (endophytic and epiphytic) induce host resistance to environmental stresses. This is commonly linked to the production of salicilate or jasmonate, or both. These two types of signalling chemicals are associated with resistance to biotic stress. Recent molecular evidence in Arabidopsis thaliana inoculated with the plant growth promoting rhizobacterium, Paenibacillus polymyxa, indicates however, that the abiotic and biotic stress-resistance are linked.

An effective plant beneficial bacterium, Pseudomonas spp. strain PsJN, was used to enhance tolerance to transplanting stress in potato. The bacterium was originally isolated as a contaminant from Glomus vesiculiferum infected onion roots. A key element in its isolation was that the bacterium (Pseudomonas sp. strain PsJN) did not grow on a standard potato culture medium in the absence of plantlets. Pre-selection of plant beneficial bacteria on plant tissue culture media is thus recommended as a critical step in the development of an in vitro co-culture system, so the bacteria do not overgrow the culture. Dipping potato nodal cuttings in a suspension of a beneficial bacterium can dramatically affect in vitro growth of plantlets. So far, our isolate, Pseudomonas sp. strain PsJN, has been the most effective plant growth promoting bacterium under in vitro conditions. It forms endophytic and epiphytic populations when co-cultured either with potato, tomato and grapevine. Thus, in clonal propagation by nodal explants, there is no need for further re-inoculation. The bacterium stimulates plant growth and induces changes that lead to better water management. It also enhances resistance to low level pathogens of potato and tomato and to Botrytis cinerea infection of grapevine.

The inoculated plantlets had a massive and wellbranched root system after four weeks in culture, and were more advanced developmentally. The stems were sturdier with more lignin deposits around the vascular system, more root hairs and more and larger leaf hairs. Stomata function of the bacterized in vitro plantlets closely resembled those of the greenhouse hardened non-bacterized transplants. In wheat seedlings exposed to osmotic stress, improvement of water relations has been observed by their co-culture with Azospirillum brasiliense strain SP245. Enhanced transplant survival of black locust, inoculated with different Rhizobium isolates selected for growth vigour from Rabinia varieties, has also been reported.

The stimulation of adventitious rooting in cultures with Agrobacterium rhizogenes is well known. It was successfully used to enhance rooting of two Mexican species of Pinus. Rooting of somatic embryo-derived adventitious shoots (15-30 mm long) increased between 7-13 per cent, and in shoots co-cultured with A. rhizogenes to 67 per cent compared with 60 per cent in the auxin treated controls. Root development was also induced in slash pine explants on co-culture with the bacterium isolates. The bacterium-induced roots resembled seedling roots rather than hairy roots, which are induced with A. rhizogenes. Induction of root elongation has been observed in potato plants co-cultured in vitro with some strains of Rhizobium leguminosarum biovar trifolii, isolated from the root nodules of red clover. The bacterization of shoot explants of oregano with a Pseudomonas sp. prevented vitrification. The oregano plantlets co-cultured with the bacterium had lower water content, but more phenolics and chlorophyll than the non-bacterized controls.

Similar responses were recorded with potato and vegetable crops. Bacterized plantlets were greener, had elevated levels of cytokinins, PAL (phenylalanine ammonia-lyase), free phenolics (especially chlorogenic and caffeic acids), and contained more lignin. In potato, elevated PAL activity in bacterized plants was found during the first four weeks of growth, there being no difference in its level in six week-old plants. However, when explants were taken from six-week-old bacterized and control stock plants, PAL activity in the newly derived shoots was higher in the pre-bacterized treatment. This suggests that pre-bacterized plantlets respond faster to cutting injuries, a critical factor in the induced host resistance to pathogen attack. In vitro bacterization of potato and vegetable plantlets significantly enhanced their post transplanting survival vigour. Enhanced vigour of potato plantlets co-cultured with isolates of Pseudomonas fluorescence isolated from tubers, during and after weaning has also been reported.

In greenhouse experiments, plants derived from dual cultures of potato and our pseudomonad bacterium had larger root systems and stolons, tuberized earlier, and gave higher tuber yields than the non-bacterized controls. The bacterized potato plantlets transplanted directly from culture vessels to the field had significantly better survival than the non-bacterized controls; however, the tuber yield varied from year to year; high precipitation or severe drought reduced the benefits of bacterization.

Tissue culture techniques provide an opportunity for the introduction of nitrogen fixing endophytes (other than Rhizobia) into clonally propagated plants for sustainable production systems. A protocol has been developed to inoculate micropropagated sugarcane plantlets with Acetobacter diazotrophicus at the end of the rooting period on medium with 10X diluted salts and sugar but without hormones and vitamins. Thirty days after transplanting, the bacteria could be re-isolated from the plant tissues. Stable artificial associations between plants and nitrogen fixing bacteria have been also reported.

A symbiotic culture system of callus and bacteria between Daucus carota L. and Azotobacter zettuovii (CRS-H6) was developed. The callus grew for four years on the nitrogen-free medium with lactose as the carbon source. The bacteria located in the intracellular spaces were transmitted to newly regenerated plantlets and fixed nitrogen. Similar stable associations were established with tomato, potato, wheat, sugarcane and poplar, using different strains of eight Azotobacter species (Varga, S.S. Personal communication), and in strawberries with Azomonas insignis. In tomato, the bacterium was transmitted through the seed. Approximately, 20 per cent of the seedlings from the seed of tomato cocultured with Azotobacter beijerinckii showed N2 fixing activity two months after germination.

Ex Vitro Bacterization

To comply with the certification requirements of plant tissue culture propagules, commercial laboratories are hesitant to introduce microorganisms into their in vitro cultures. One approach is to wean plants in rhizosphere with both beneficial bacteria and mycorrhizal fungi. In potato, the tissue-cultured plantlets are readily infected both in external and internal tissues with bacteria present in the transplanting medium. This approach, although attractive, will have to deal with the problem of maintaining stable populations of the introduced microorganisms. The beneficial effects of bacterial inoculation on tree seedlings and greenhouse produced potato tubers show the potential of post-vitro bacterization of tissue-cultured propagules. The post-vitro mycorrhization and bacterization of micropropagated strawberry, potato and azalea with certain combinations of bacteria and mycorrhiza enhanced greenhouse production of minitubers, and a mixture of three strains of rhizobacteria improved the post-transplanting performance of strawberries.

Mycorrhization

The inoculation of tissue-cultured plantlets with endomycorrhizal fungi is beneficial to plant growth. It is well known that the root colonization with vesicular-arbuscular mycorrhizae (VAM) improves plant nutritional status, water management and disease resistance. Several secondary metabolites (cyclohexenone derivatives) with structural similarity to abscisic acid have been identified in tobacco roots in response to the Glomus intraradices infection. A variety of phenylpropanoid compounds induced by ectomycorrhizal inoculants, have also been identified in several conifers. Such compounds are related to the biotic and abiotic stress management in plants. To improve the performance of tissue-cultured plantlets, rooted-shoots have been inoculated with mycorrhiza at transplanting. The benefits of mycorrhization depended on the growing medium, plant and mycorrhizal species, and the degree of root colonization. Because VAM stimulated growth of plants was frequently expressed only after acclimatization, it was suggested to develop a culture system where the mycorrhizal fungi are introduced in vitro during the rooting stage.

It is possible to grow fungi in vitro on Ri T-DNA transformed carrot roots under elevated CO_2. Capitalizing on this technique, a 'tripartite' culture system for in vitro inoculation of strawberries and vegetable crops was developed. The most effective treatment consisted of 30 day-old VAM transformed carrot-root culture and strawberry shoots in cellulose plugs. After root induction, the plugs were placed on the surface of the mycorrhized root culture and kept in a growth chamber under 5000 PPM CO_2 for 20 days.

All plantlets were successfully colonized, and exhibited a larger root system, better shoot growth, and a higher (more negative) osmotic potential when compared to non-mycorrhized controls. It is suggested that the enhancement of osmotic potential is important in the pre-adaptation step prior to full acclimation of plantlets for transplanting.

The high osmotic potential of wheat seedlings co-cultured with Azospirillum allowed them to withstand osmotic stress much better than non-inoculated controls with low osmotic potential. In strawberry, mycorrhized plants had a better establishment rate and produced more runners than non-mycorrhized controls. The plants were grown in polyurethane foam substrate under photoautotrophic conditions with reduced sucrose concentration in the medium. The foam tear-away strips fit into any culture vessel, and the VAM spores can be placed directly in the planting holes. In garlic, improved growth was observed after post-vitro transplant inoculation. Increased rooting and reduction in weaning stress have been reported in the medicinal plant, Babtista tinctoria (L.) R. BR..

Several studies have demonstrated synergistic effects of VAM fungi and diazotrophic bacteria on nutrition and growth of various crops. Such effects of "mycorrhiza helper bacteria" need to be explored in tissue culture systems. Creation of defined tissue culture micro-ecosystems could allow the study of complexed plant–microbial-environment interactions, which could refine and improve our traditional in vitro propagation methods and prime the propagules for ex-vitro environments.

Priming with Microbial Derivatives

Cyanobacterial and fungal culture extracts have been shown to induce developmental changes and pathogen resistance in tissue-cultured material when added to the medium. Microbial culture filtrates have been reported for the control of plant tissue culture contaminants. An acetone precipitated fraction of Bacillus subtilis Ehrnberg and Trichoderma viride Pers exhibited high antifungal activity, and extracts from Pseudomonas fluorescence Migula (strain X) had high antibacterial activity. The bacterial extracts did not reduce the performance of cultures of Nicotiana tabacum L. over four subcultures but reduced the growth of accidental contaminants. Heat-stable mycelial extracts of another Trichoderma species, a non-pathogenic T. longibrachiatum were triggered induced-resistance response to Phytophthora parasitica var. nicotianae (race 0) in tobacco seedlings. The extract induced expression of pathogenesis-related proteins, PR-16 and osmotin (PR-5) at a higher level than the extract prepared in the same manner from the pathogenic fungus.

Incorporation of Low Cost Options

For many plants, the conventional methods of propagation are cheaper than tissue-culture technology. Low cost options can generally be incorporated into the design of the building, laboratories, working areas, layout of equipment, lighting, heating and production planning to provide smooth and efficient operations.

It is important to select several plants that provide options for production around the year to allow cash flow and optimal use of equipment and facilities. It is essential to maintain sufficient mother cultures, and limit the number of subcultures to avoid variation, and plan production and shipment according to the customer's demand. Low cost options can generally be incorporated into the design of the building, laboratories, working areas, layout of equipment, lighting, and heating to provide smooth and efficient operations. However, some options are stage-specific to the process of micropropagation and others are linked to marketing. The following strategies should be considered before venturing into commercial micropropagation.

Selection of Crops

Micropropagation protocols have been developed for a large number of species; however, only a limited number of plants are being produced on a large scale through micropropagation. This is because conventional methods of propagation are cheaper than tissue culture technology or there is a selective market demand. Ornamentals account for approximately 80 per cent of the world trade in the tissue culture industry. Plants produced by commercial laboratories should yield steady revenue, *e.g.* Spathiphyllum, Syngonium, lilies. Some plants are unique to individual commercial companies.

The life cycle of the species and the cropping system adopted by growers are important criteria to select plants for large-scale tissue culture. Initially, micropropagation units focused on the production of ornamental plants that were least affected by the season, *e.g.* foliage and indoor plants. But to capture larger markets, commercial tissue culture units have ventured into fruits, vegetables, landscape and medicinal plants, which provide options for production around the year. This leads to improved cash flow and optimal use of equipment and facilities. When selecting specific varieties protected under Plant Breeders Rights, these should be multiplied only with the agreement of the breeders.

Target Markets

Before going commercial, one must ensure that there is a large and expanding market for the product. Both the domestic and export markets should be explored extensively. It must be ensured that large contracts are available each year. Products that command a high price should be preferably chosen.

Regularity of Dispatches

It is essential to plan the production schedule according to the customer's demand and despatch the products at the stipulated time. It is not in the interest of the tissue culture facility to remain idle. Production of different plants should be distributed over the year to make optimum use of the laboratory facilities. Regularity of despatches adds to the viability of the business.

End Product

It is essential to decide on the type of products to be produced in a tissue culture facility. These could be plants that are rooted or non-rooted and could be sold in agar or marketed as hardened propagules in plugs or trays. Relatively small tissue culture units can produce nonrooted microcuttings in large numbers. Rooted microcuttings on the other hand require large growth room areas. In both cases, dispatching them ex-agar may result in some mortality at the grower's end. To reduce mortality, it is important to ship them in-agar. However, in-agar shipments occupy a large volume, and hence require more cargo space and add to the cost. Such high transport and cargo charges can only be justified for high value products. The hardened micropropagated plants in plugs or trays are the easiest to handle by growers.

Prevention of Variation in Tissue Culture

The establishment of mother cultures from explants is a significant part of the production cost. Production projections can get out of gear due to the lack of sufficient mother cultures and maintenance of large numbers of such cultures adds to the cost of production. There is a general consensus that the number of subcultures for explants should be limited to avoid somaclonal variation. For example, in potato, the number of subcultures is recommended to be around twenty. Hence, repeated explant culture initiation becomes necessary to have sufficient number of mother cultures whenever largescale production is planned. For many plants, it is not known whether limiting the number of subcultures reduces genetic variation. Many of the plants that are propagated from vegetative parts such as banana, potato, sweet potato, sugarcane, chrysanthemum and many fruits are polyploids and well buffered against the expression of mutations and somaclonal variation. If axillary bud proliferation is used as the strategy of multiplication, the frequency of variation is low.

If variation does occur, the number of variants will rise with increasing number of subcultures unless there is selective elimination in tissue culture. The high frequency of somaclonal variation has prevented the use of tissue culture for mass scale propagation of oil palm. In general, the initiation of cultures from meristem-tip gives low variation in subcultures. Some of the novel somaclonal variants can be of value in expanding the market.

Multiplication and Hardening

Liquid media reduce the costs of multiplication. Increasing the multiplication rate through enhanced axillary bud proliferation helps in cutting cost of production. Mechanical cutting and bioreactors is a valuable option at Stage II. In some systems, shoot elongation and rooting are two separate phases. Combining them in one step by using appropriate media with auxins and cytokinins is a valid cost cutting strategy. Reducing the duration in the hardening facility but increasing the survival rate reduces production costs. If natural light is used during multiplication and rooting, the plants need relatively shorter duration for hardening. Acclimatization in vitro helps in better survival when the plants are transferred to soil. Loss of plants at the hardening stage contributes to increased costs of production. A cost-effective system based on natural sunlight for in vitro acclimatization gives high survival. If grown under artificial lighting, tissue cultured plants should be kept under shade for some time and only then subjected to full sun hardening.

Greenhouse Facility

Sophisticated and costly greenhouses can be substituted by cost-effective techniques for hardening in vitro derived plantlets. For this, it is necessary to understand the changes that tissue-cultured plantlets go through during the transfer from in vitro to in vivo conditions.

Micropropagated plantlets should be acclimatized gradually by transferring them to a clean area under partial shade. In a low-cost hardening system, the culture containers are kept in the greenhouse with lids loosened. The plants are allowed to grow for 4-6 weeks under conventional Grow lights TM. This semi-hardens plants, and leads to shoot elongation, which is usually done under laboratory conditions. Contamination can be controlled with regular sprays of fungicides and bactericides.

Table: *Changes in the Environment for Hardening in Vitro Plants to Ex-vitro Conditions*

In vitro conditions	*Ex-vitro conditions*
Temperature 25 ± 2°C	Temperature 23-36°C
Light intensity 1200-2000 lux	Light intensity 4000-12000 lux
Narrow spectrum light	Broad spectrum light
Relative humidity 98-100%	Relative humidity 40-80%
Almost non-functional roots	Functional roots
Almost non-functional photo - synthetic system	Functional photosynthetic - system
Exogenous hormones	Endogenous hormones
Sterile conditions	Non-sterile conditions

The plants are periodically sprayed with water to maintain humidity. Bench lamps are used to provide light, if and when required. Semihardened elongated shoot clumps are then separated under non-sterile conditions, and rooted in peat or plugs under thin plastic sheets for 2-3 weeks. This low-cost alternative reduces the capital cost of hardening by saving laboratory space, labour, and time. This procedure has been used on a commercial scale for a number of plants such as Gerbera, Spathiphyllum, Syngonium, African violets, Chrysanthemum, Ficus, Philodendron and other ornamentals in the Netherlands and Israel. The growers in Florida harden tissue-cultured foliage plants in cavity trays covered with a transparent plastic tray, thus saving costs of a greenhouse. In some parts of India, hanging the culture plastic bags in shade hardens tissue-cultured sugarcane plants. Plants produced in this manner are semihardened, and can be delivered to the growers at a relatively cheaper price.

Low-cost plastic tunnels (polytunnel) with bio-fertilization can also be used for hardening, especially in cooler climates. A rectangular pit of desired and manageable size is dug on a site free of water logging. A frame of bamboo or any other available material is made above the pit with a gentle slope to one side. The frame is then covered with a clear polythene sheet, and tied using nylon ropes. The sheet is sealed with mud on three sides, leaving one side free.

The temperature and humidity inside these structures is generally higher than the outside, and protects the plants from frost. The polythene sheet can be partly opened to allow air circulation and sunlight as and when required. The added advantage of polypits is the carbon dioxide fertilization effect and reduced need for watering. The International Institute for Tropical Agriculture, Ibadan, Nigeria has developed another low cost hardening alternative. The institute provides the farmers with plants in plastic bags and handling instructions. For hardening the plants, the farmers hang the plastic bags under tree shade.

Integration of Technology from Lab to Land

The price of the tissue culture derived plants, tubers, bulbs, and cuttings must be competitive with those obtained from conventional propagation. Many tissue-cultured plants are too expensive for direct field planting. In such cases, the cost can be reduced by one or more times conventional propagation of the tissue cultured plants. The uniformity and consistency in field performance of tissue-cultured plants is important to build confidence of the farmers to integrate such plants in the production systems

Tissue culture technology contributes significantly to the improvement of agricultural productivity and food security. However, it must be integrated into production systems in a cost effective manner. Wherever practical, direct use of the tissue-cultured plants is the best strategy to obtain the full benefits of this technology. The price of the products, *i.e.* plants, tubers, bulbs and cuttings, must be competitive with those obtained from conventional propagation. While many plants that are propagated from vegetative parts can be micropropagated, the cost of tissue-cultured plants is still too high for direct growing by the farmers. For example, micro- and mini-potato tubers and micropropagated plants of chrysanthemum, strawberry, sugarcane and cassava are too expensive for direct field planting. In such cases, tissue-cultured plants are grown as super-elite or elite material and the costs are brought down by conventional propagation one or more times. Thus, the costly tissue-cultured plants are used as nuclear material, and their further multiplication is by conventional vegetative propagation in the field to produce large quantities of pathogen-free planting material.

However, the subsequent vegetative multiplication of tissue culture-derived material in the open environment also exposes them to infection with diseases and viruses that may increase in successive propagation. In addition, the handling of bulky conventional planting material, such as potato tubers, banana corms and sugarcane sets adds to the transportation costs. Hence, the cost of tissue-cultured plants has to be brought down substantially, so that farmers benefit through increased productivity.

On-Farm Technology Transfer

The most outstanding application of tissue culture in crops that are propagated from vegetative parts has been the production of disease-free planting material that gives improved yield and quality. This has been particularly the case in potato, banana, citrus, date palm, pineapple, papaya and many ornamentals. Seed potato production is a highly specialized business, and requires highly skilled growers. Currently, all the certified seed potato in Ireland and Scotland is derived from tissue-cultured plants, which produce micro- or mini-tubers that form the super-elite seed. The mini-seed tubers are not used for growing the potato crop. Instead, these are further multiplied by conventional propagation in the field, at least twice, and sometimes even 3 to 4 times for the production of certified seed tubers.

Apart from cost reduction and high seed quality, the farmers get the same types of seed tubers with which they are used to. In Canada, many growers have integrated small tissue-culture laboratories for seed potato production on their farms. Similarly, the sugarcane growers in Punjab, India have switched to high quality seed cane that is derived from tissue cultured plants, which are propagated 3 to 4 times from conventional cuttings. The lead in commercial sugarcane tissue culture technology has been taken by the sugar cane mills, which have established their own laboratories for large-scale seed cane production. The state agricultural university provides the disease-free indexed cultures to the sugar mills. Such a model that includes the university for research and trouble-shooting problems, private industry (sugar mills) that multiplies the sugarcane seed material through tissue culture, and cane growers who produce the crop, has been very successful in integrating tissue culture technology for the benefit of the farmers and processors.

This tripartite cooperation has rapidly replaced conventional sugarcane seed production in the state. Recently, direct planting of micropropagated sugarcane plants has been profitably undertaken in Maharashtra, India. In Morocco, citrus indexed scions are multiplied in tissue culture and grown under a plastic house. Scion cuttings with two-three buds each are then sold to the farmers, who graft them on either new or old rootstock. This has helped to eradicate many of the citrus viruses and yet retain the old rootstock plants that are resistant to diseases. In another approach, high quality disease-free tissue-cultured scions are grafted in vitro onto seedling rootstock. Millions of such in vitro grafted plants have been distributed to the citrus growers in Spain, bringing direct benefits to farmers.

In Thailand, tissue-cultured chrysanthemum plants, produced by the Department of Agriculture, are sold to farmers who multiply them further from conventional cuttings for the production of cut flowers. This procedure has allowed rapid introduction of new varieties to respond to the market demand. In Vietnam and China, in vitro cultured plants of sweet potato and cassava are distributed directly to the farmers.

The farmers multiply in vitro plants further from conventional cuttings, and produce sufficient amounts of healthy planting material for their on- farm use. A similar procedure could be used for the distribution of disease-free material of cassava in Africa where cassava mosaic virus disease has become a major threat to cassava production. Perhaps, the most successful on-farm use of tissue culture technology has been in Cuba where an ordinary village-house is converted into a plant tissue culture facility – a 'biofactory'. Such a facility employs local labour, particularly housewives, for various operations, uses low cost media and containers and sunlight, and has the capacity to produce one million banana or sugarcane plants per year.

In banana, the hardened plants are sold to the local banana growers. A batch of one thousand plants is kept in the nursery to monitor the quality. The farmers are asked to notify any off-type, which they may observe. This has not only given the farmers a confidence in the technology but has also resulted in the discovery of several improved variants among the tissue culture derived populations.

In developed countries, many commercial companies restrict tissue-culture propagation to the production of in vitro plants, which are used as nucleus or 'super elite' material, and are further multiplied two to three times before being handed over to the farmers as certified material. Other companies subcontract the production of fully-grown plants on a buy-back basis for further sale. This approach has been used in the propagation of many ornamental plants and shrubs in Belgium.

In the developing countries, tissue culture must be seen as a complementary approach to conventional plant propagation, so that the rural job opportunities are not lost, but strengthened and further expanded. Hence, on-farm conventional multiplication of the tissue culture propagated material is the cheapest and the most sustainable method for cost reduction, and for the creation of farmer confidence in the technology.

Of course, the agricultural extension workers and propagation companies must provide the necessary know-how to the farmers for handling and growing tissue cultured plants for further multiplication under high health status. Universities and experts should provide training to farmers and extension workers as being done in several countries.

On-Farm Performance

Even after a decade of introduction of millions of tissue culture plants, the market covered is less than 1 per cent of the total cultivation under vegetatively propagated crops. Thus, there is a tremendous scope to use tissue-culture technology to provide growers with nuclear stocks of shrubs, fruit and forest trees. There is a need to micropropagate hundreds of native varieties and local cultivars of banana and plantain, cassava, citrus, date palm, ginger, potato, sweet potato, turmeric, yams, and local tropical fruits, flowers, herbs, medicinal plants and trees that are propagated from vegetative parts.

The consistency in field performance of tissue-cultured plants is extremely important. Farmer's participation is essential in achieving this goal. Precision agricultural management, based on integrated and sustainable crop nutrient practices, is required to obtain high yields from quality planting material.

In some cases, the problem of unsatisfactory field performance was traced to nutritional imbalance and poor soil conditions. The commercial micropropagation companies should encourage the availability and use of site-specific agronomic practices by the growers to ensure sustainable success. This would provide farmers the latest technology product with minimal risk. Thus, to achieve expansion of commercial micropropagation, the use of cost-effective technology and of locally adapted varieties and relevant agronomic practices is necessary.

Plant tissue culture refers to growing and multiplication of cells, tissues and organs of plants on defined solid or liquid media under aseptic and controlled environment. Micropropagation allows rapid production of high quality, disease-free and uniform planting material. The micropropagation of high quality planting material of ornamentals, and forest and fruit trees has created new opportunities in global trading for producers, farmers, and nursery owners, and for rural employment. The plants can be multiplied under a controlled environment anywhere irrespective of the season and weather on a year-round basis. However, micropropagation technology is more expensive than the conventional method of plant propagation. It is a capital-intensive industry, and in some cases the unit cost per plant becomes unaffordable. Hence, it is necessary to adopt strategies to reduce production cost and lower the cost per plant.

Plant micropropagation is primarily based on rapid proliferation of tiny stem cuttings, axillary buds, and to a limited extent of somatic embryos, cell clumps in suspension cultures and bioreactors. The cultured cells and tissue can take several pathways. The pathways that lead to the production of true-to-type plants in large numbers are the preferred ones for commercial multiplication. The process of micro-propagation is usually divided into several stages *i.e.*, pre-propagation, initiation of explants, subculture of explants for proliferation, shooting and rooting, and hardening. These stages are universally applicable in large-scale multiplication of plants. The delivery of hardened small micropropagated plants to growers and market also requires extra care.

Low-cost tissue culture technology is the adoption of practices and use of equipment to reduce the unit cost of micropropagule and plant production. Low cost options should lower the cost of production without compromising the quality of the micropropagules and plants. In low cost technology cost reduction is achieved by improving process efficiency and better utilization of resources. Low-cost tissue-culture technology is high priority in agriculture, horticulture, forestry, and floriculture of many developing countries for the production of suitably priced high quality planting material.

A number of low-cost alternatives can be used to simplify various operations and reduce the costs in a tissue culture facility. The physical components of a typical plant tissue culture facility include equipment and buildings with preparation room, transfer room, culture or growth room, hardening and weaning area, soil-growing area (greenhouses, plastic tunnels), packaging and shipping area, and related facilities such as an office, and a store for chemicals, containers and supplies. The size of the physical components of a tissue culture facility will vary according to its functional needs, *i.e.* the volume of production. Careful planning of a facility can make large savings both in the construction costs and day-to day operations in the facility. It is recommended that an existing facility should be visited to view the layout and operational needs before starting a new facility.

Proper choice of media and containers can reduce the cost of micropropagation. The composition of culture media used for proliferation has a tremendous influence on production costs. The type of culture vessel influences the efficiency of transfer during subculture and production of propagules per unit area. The replacement of expensive imported vessels with reusable glass jars and lids, alternatives to gelling agents, use of household sucrose, and some medium components can reduce costs of production. Bulk making of media and storage as deep frozen stocks also reduces labour costs.

Artificial lighting of cultures in the growth rooms is one of the most expensive and inefficient methods in tissue culture technology. Changing the method of illumination from artificial to natural light is a decisive low cost option in tissue culture. This reduces electricity and capital costs and also improves the plant quality. Maintaining in vitro cultures at a regulated temperature with air conditioners adds to the cost but does not contribute to specific plant quality. Many in vitro growing plants can tolerate wide fluctuations in temperature, and adapt better to field conditions than those grown under even temperature. Plants can also be hardened in open shade. Plants hardened under natural light are sturdy, and withstand transplantation better in the field. Production of plants based on tissue culture technology and their subsequent growing is a labour intensive system. Even in developing countries, where labour is relatively less costly, hiring expertise from established R&D laboratories reduces overhead costs.

Bioreactor-based propagation of plants can increase rate of multiplication and growth of cultures and reduce space, energy and labour requirements in commercial micropro-pagation. They can therefore be attractive to developing countries as regards new or expanding plant culture facilities, in combination with a conventional laboratory. Bioreactors provide more precise control of the plant growth gaseous exchange, illumination, medium agitation, temperature and pH than the conventional culture vessels. However, to be cost-effective, use of bioreactors requires indexed plant cultures, and attention to aseptic procedures during handling of plant material otherwise culture contamination leads to massive economic loss. In micropropagation, the health status of the donor mother plant and of the plants multiplied from it are among the most critical factors, which determine the success of a tissue culture operation.

The indexing of the mother plants for freedom from viral, bacterial, and fungal diseases is a normal procedure in large-scale plant propagation through tissue culture. Plants not originating from pathogen-tested material must be screened for the presence of viruses. Laboratories, which do not have in-house facilities to carry out plant indexing, should obtain their indexed stock plants from organizations such as Departments of Agriculture, agricultural universities or privately owned certified germplasm repositories that routinely produce such plant material.

Batches of micropropagated plants should be tested for freedom from diseases either in-house or by other laboratories. ELISA has been the most effective method for virus and pathogen detection in plants. Polymerase chain reaction and nucleic acid hybridization are more sensitive than ELISA, and can detect pathogens in extremely low amounts.

Quality control is essential to assure production of high quality plants and to have end-users confidence. Quality standards require the establishment of suitable tests to maintain quality control. The choice of explant source, freedom of the donor plant from viruses, disease causing fungi, bacteria, viroids, phytoplasmas, vigour and conformity of the variety, and elimination of somaclones are critical for maintaining plant quality.

Variety identification by proper labeling at all stages is essential to ensure varietal identity. Tissue-cultured propagules are produced under a controlled environment. Such plantlets have small juvenile leaves with reduced photosynthetic capacity, and malfunctioning stomata. Priming for rooting, shooting, and improved photosynthesis can be achieved with growth regulators and adjustment to the growing conditions that affect the post-transplanting performance of the propagules. Vented closures with microbial filters facilitate gas exchange, reduce ethylene build-up that stunts plant growth, reduces leaf-size, and causes leaf drop in tissue culture containers. Plantlets produced under photo-autotrophic culture systems on media with or without sucrose but CO_2-enrichment, increased light intensity, good gas exchange and reduced humidity are more vigourous, have larger root-systems, and are less susceptible to microbial contamination. Plants adapted gradually to the ex-vitro environment have improved survival upon transfer to soil. Plants in their natural environment are colonized with many bacteria, fungi, and mycorrhizae. In-vitro or ex-vitro biopriming of micropropagated plants with such organisms improves plant performance under stress environments, and consequently enhances yield.

Low cost options can generally be incorporated into the design of the building, laboratories, working areas, layout of equipment, lighting, heating and production planning to provide smooth and efficient operations. It is important to select several plants that provide options for production around the year to allow cash flow and optimal use of equipment and facilities. It is essential to maintain sufficient mother cultures, and limit the number of subcultures to avoid variation, and plan production and shipment according to the customer's demand. The price of the tissue culture derived plants, tubers, bulbs, and cuttings must be competitive with those obtained from conventional propagation. Many tissue-cultured plants are too expensive for direct field planting. In such cases, the cost can be reduced by one or more conventional propagations of the in vitro plants. The uniformity and consistency in field performance of tissue-cultured plants is important to build confidence of the farmers to integrate such plants in the production systems.

Transvaginal Oocyte Retrieval

Transvaginal oocyte retrieval (TVOR), also referred to as oocyte retrieval (OCR) or even simply egg collection, is a technique used in vitro fertilization (IVF) in order to remove oocytes from the ovary of the female, enabling fertilization outside the body. Transvaginal oocyte retrieval can be called transvaginal ovum retrieval if the oocytes have matured into ova, as is normally the case in IVF.

TVOR is performed in an operating room with the (female) patient in the lithotomy position. Initially performed using a transvaginal needle guided by transabdominal ultrasonography, TVOR is currently performed with a transvaginal ultrasound transducer with an attached needle. Once the follicle of the ovary is lined up with the guiding line of the transducer, the needle is advanced through the vaginal wall into the follicle of the ovary.

Care is taken not to injure organs located between the vaginal wall and the ovary. Once the follicle is entered, suction is gently applied to aspirate follicular fluid and with it, hopefully, cellular material including the oocyte. Next, other follicles are aspirated.

Once the ovarian follicles have been aspirated on one ovary, the needle is withdrawn, and the procedure repeated on the contralateral ovary. After completion, the needle is withdrawn, and hemostasis is achieved. The process takes about 20 minutes and is usually done under procedural sedation or general anesthesia.

The technique mostly succeeds after ovarian hyperstimulation, where oocytes are pharmacologically stimulated to mature. In addition, oocyte release triggering is usually performed to induce final maturation of oocytes. An agent used for this purpose is human chorionic gonadotropin (hCG). It would cause ovulation to occur between 38 and 40 hours after injection, but TVOR is typically performed at a time usually between 34 and 36 hours after hCG injection, that is, just prior to the expected rupture of the follicles. hCG injection confers a risk of ovarian hyperstimulation syndrome.

The retrieval can lead to complication resulting in injury to organs, bleeding, and infection. The aspirates are handed to the laboratory to examine it for oocytes. Among lean patients with polycystic ovary syndrome, ovarian hemorrhage after ultrasonographically guided transvaginal oocyte retrieval is a potentially catastrophic and not so rare complication.

Anecdotal evidence suggests that certain airborne chemical contaminants and particles, especially volatile organic compounds (VOC), may be toxic to and impair the growth and development of embryos if present in sufficient concentrations in the ambient atmosphere of an IVF incubator. This technique was first developed by Pierre Dellenbach and colleagues in Strasbourg, France, in 1984 and introduced into the United States at the Genetics and IVF Institute in Virginia in 1985 with the help of Deborah Costakos.

Egg and Sperm Preparation

In the laboratory, the identified eggs are stripped of surrounding cells and prepared for fertilization. An oocyte selection may be performed prior to fertilization to select eggs with optimal chances of successful pregnancy.

In the meantime, semen is prepared for fertilization by removing inactive cells and seminal fluid in a process called sperm washing. If semen is being provided by a sperm donor, it will usually have been prepared for treatment before being frozen and quarantined, and it will be thawed ready for use.

Fertilization

The sperm and the egg are incubated together at a ratio of about 75,000:1 in the culture media for about 18 hours. In most cases, the egg will be fertilized by that time and the fertilized egg will show two pronuclei. In certain situations, such as low sperm count or motility, a single sperm may be injected directly into the egg using intracytoplasmic sperm injection (ICSI). The fertilized egg is passed to a special growth medium and left for about 48 hours until the egg consists of six to eight cells.

In gamete intrafallopian transfer, eggs are removed from the woman and placed in one of the fallopian tubes, along with the man's sperm. This allows fertilization to take place inside the woman's body. Therefore, this variation is actually an *in vivo* fertilization, not an *in vitro* fertilization.

Embryo Culture

Typically, embryos are cultured until having reached the 6–8 cell stage three days after retrieval. In many Canadian, American and Australian programmes, however, embryos are placed into an extended culture system with a transfer done at the blastocyst stage at around five days after retrieval, especially if many good-quality embryos are still available on day 3.

Blastocyst stage transfers have been shown to result in higher pregnancy rates. In Europe, transfers after 2 days are common. Culture of embryos can either be performed in an artificial culture medium or in an autologous endometrial coculture (on top of a layer of cells from the woman's own uterine lining).

With artificial culture medium, there can either be the same culture medium throughout the period, or a *sequential system* can be used, in which the embryo is sequentially placed in different media. For example, when culturing to the blastocyst stage, one medium may be used for culture to day 3, and a second medium is used for culture thereafter.

Single or sequential medium are equally effective for the culture of human embryos to the blastocyst stage. Artificial embryo culture media basically contain glucose, pyruvate, and energy-providing components, but addition of amino acids, nucleotides, vitamins, and cholesterol improve the performance of embryonic growth and development.

Methods to permit dynamic embryo culture with fluid flow and embryo movement are also available. A new method in development uses the uterus as an incubator and the naturally occurring intrauterine fluids as culture medium by encapsulating the embryos in permeable intrauterine vessel.

Embryo Selection

Laboratories have developed grading methods to judge oocyte and embryo quality. In order to optimise pregnancy rates, there is significant evidence that a morphological scoring system is the best strategy for the selection of embryos. However, presence of soluble HLA-G might be considered as a second parameter if a choice has to be made between embryos of morphologically equal quality.

Also, two-pronuclear zygotes (2PN) transitioning through 1PN or 3PN states tend to develop into poorer-quality embryos than those that constantly remain 2PN. More advanced methods of *embryo profiling* may also be performed in order to optimise embryo selection, as further described in the "expansions"-section below.

Embryo Transfer

Embryos are failed by the embryologist based on the amount of cells, evenness of growth and degree of fragmentation. The number to be transferred depends on the number available, the age of the woman and other health and diagnostic factors.

In countries such as Canada, the UK, Australia and New Zealand, a maximum of two embryos are transferred except in unusual circumstances. In the UK and according to HFEA regulations, a woman over 40 may have up to three embryos transferred, whereas in the USA, younger women may have many embryos transferred based on individual fertility diagnosis.

Most clinics and country regulatory bodies seek to minimise the risk of pregnancies carrying multiples. As it is not uncommon for more implantations to take than desired, the next step faced by the expectant mother is that of selective abortion. The embryos judged to be the "best" are transferred to the patient's uterus through a thin, plastic catheter, which goes through her vagina and cervix. Several embryos may be passed into the uterus to improve chances of implantation and pregnancy.

Embryos can be either "fresh" from fertilized egg cells of the same menstrual cycle, or "frozen", that is they have been generated in a preceding cycle and undergone embryo cryopreservation, and are thawed just prior to the transfer. The outcome from using cryopreserved embryos has uniformly been positive with no increase in birth defects or development abnormalities, also between fresh versus frozen eggs used for intracytoplasmic sperm injection (ICSI). Children born from vitrified blastocysts have significantly higher birthweight than those born from non-frozen blastocysts.

Uterine Preparation

In the human, the uterine lining (endometrium) needs to be appropriately prepared so that the embryo(s) can implant. In a natural or stimulated cycle, the embryo transfer takes place in the luteal phase at a time where the lining is appropriately undeveloped in relation to the status of the present Luteinizing Hormone. In a cycle where a "frozen" embryo is transferred, the recipient woman could be given first estrogen preparations (about 2 weeks), then a combination of oestrogen and progesterone so that the lining becomes receptive for the embryo. The time of receptivity is the implantation window.

Limited evidence also supports removal of cervical mucus before transfer.

In stimulated cycles in human IVF, embryos are typically transferred 3 days after fertilization and may then be at the eight-cell stage, or they are transferred 2 to 3 days later when they have reached the blastocyst stage. Embryos who reach the day 3 cell stage can be tested for chromosal or specific genetic defects prior to possible transfer by preimplantation genetic diagnosis (PGD).

Monozygotic twinning is not increased after blastocyst transfer compared with cleavage-stage embryo transfer.

The embryo transfer procedure starts by placing a speculum in the vagina to visualise the cervix, which is cleansed with saline solution or culture media. A soft transfer catheter is loaded with the embryos and handed to the clinician after confirmation of the patient's identity. The catheter is inserted through the cervical canal and advanced into the uterine cavity.

There is good and consistent evidence of benefit in *ultrasound guidance*, that is, making an abdominal ultrasound to ensure correct placement, which is 1–2 cm from the uterine fundus. Anesthesia is generally not required.

Single embryo transfers in particular require accuracy and precision in placement within the uterine cavity. The optimal target for embryo placement, known as the maximal implantation potential (MIP) point, is identified using 3D/4D ultrasound. However, there is limited evidence that supports deposition of embryos in the midportion of the uterus.

After insertion of the catheter, the contents are expelled and the embryos are deposited. Limited evidence supports making trial transfers before performing the procedure with embryos. After expulsion, the duration that the catheter remains inside the uterus has no effect on pregnancy rates. Limited evidence suggests avoiding negative pressure from the catheter after expulsion. After withdrawal, the catheter is handed to the embryologist, who inspects it for retained embryos.

In the process of zygote intrafallopian transfer (ZIFT), eggs are removed from the woman, fertilized, and then placed in the woman's fallopian tubes rather than the uterus.

Embryo Number

A major issue is how many embryos should be transferred. Placement of multiple embryos carries the risk of multiple pregnancy. In the past, physicians have often placed too many embryos in the hope to establish a pregnancy. However, the rise in multiple pregnancies has led to a reassessment of this approach. Professional societies and in many countries, the legislature, have issued guidelines or laws to curtail a practice of placing too many embryos in an attempt to reduce multiple pregnancies.

The technique of selecting only one embryo to transfer to the woman is called elective-Single Embryo Transfer (e-SET) or, when embryos are at the blastocyst stage, it can also be called *elective single blastocyst transfer (eSBT)*. It lowers the risk of multiple pregnancies, compared with e.g. Double Embryo Transfer (DET) or *double blastocyst transfer* (2BT), with a twinning rate of approximately 3.5% in sET compared with approximately 38% in DET, or 2% in eSBT compared with approximately 25% in 2BT.

At the same time, pregnancy rates is not significantly less with eSBT than with 2BT. Furthermore, SET has better outcomes in terms of mean gestational age at delivery, mode of delivery, birthweight, and risk of neonatal intensive care unit necessity than DET. e-SET of embryos at the cleavage stage reduces the likelihood of live birth by 38% and multiple birth by 94%. Evidence from randomised, controlled trials suggests that increasing the number of e-SET attempts (fresh and/or frozen) results in a cumulative live birth rate similar to that of DET.

The usage of single embryo transfer is highest in Sweden (69.4%), but as low as 2.8% in the USA. Access to public funding for ART, availability of good cryopreservation facilities and legislation appear to be the most important factors for regional usage of single embryo transfer. Also, personal choice plays a significant role as many subfertile couples have a strong preference for twins.

Patients usually start progesterone medication after egg (also called oocyte) retrieval. While daily intramuscular injections of progesterone-in-oil (PIO) have been the standard route of administration, PIO injections are not FDA-approved for use in pregnancy.

A recent meta-analysis showed that the intravaginal route with an appropriate dose and dosing frequency is equivalent to daily intramuscular injections. In addition, a recent case-matched study comparing vaginal progesterone with PIO injections showed that live birth rates were nearly identical with both methods. A duration of progesterone administration of 11 days results in almost the same birth rates as longer durations.

Patients are also given estrogen medication in some cases after the embryo transfer. Pregnancy testing is done typically two weeks after egg retrieval.

Third-party Reproduction

It is not necessary that the embryo transfer be performed on the female who provided the eggs. Thus another female whose uterus is appropriately prepared can receive the embryo and become pregnant. Embryo transfer may be used where a woman who has eggs but no uterus and wants to have a biological baby; she would require the help of a gestational carrier or surrogate to carry the pregnancy. Also, a woman who has no eggs but a uterus may resort to egg donor IVF, in which case another woman would provide eggs for fertilization and the resulting embryos are placed into the uterus of the patient. Fertilization may be performed using the woman's partner's sperm or by using donor sperm.

'Spare' embryos which are created for another couple undergoing IVF treatment but which are then surplus to that couple's needs may also be transferred. Embryos may be specifically created by using eggs and sperm from donors and these can then be transferred into the uterus of another woman. A surrogate may carry a baby produced by embryo transfer for another couple, even though neither she nor the 'commissioning' couple is biologically related to the child.

Third party reproduction is controversial and regulated in many countries. Persons entering gestational surrogacy arrangements must make sense of an entirely new type of relationship that does not fit any of the traditional scripts we use to categorise relations as kinship, friendship, romantic partnership or market relations. Surrogates have the experience of carrying a baby that they conceptualise as not of their own kin, while intended mothers have the experience of waiting through nine months of pregnancy and transitioning to motherhood from outside of the pregnant body. This can lead to new conceptualisations of body and self.

History

The first transfer of an embryo from one human to another resulting in pregnancy was reported in July 1983 and subsequently led to the announcement of the first human birth February 3, 1984. This procedure was performed at the Harbor UCLA Medical Centre under the direction of Dr. John Buster and the University of California at Los Angeles School of Medicine. In the procedure, an embryo that was just beginning to develop was transferred from one woman in whom it had been conceived by artificial insemination to another woman who gave birth to the infant 38 weeks later. The sperm used in the artificial insemination came from the husband of the woman who bore the baby.

This scientific breakthrough established standards and became an agent of change for women suffering from the afflictions of infertility and for women who did not want to pass on genetic disorders to their children. Donor embryo transfer has given women a mechanism to become pregnant and give birth to a child that will contain their husband's genetic makeup.

Although donor embryo transfer as practiced today has evolved from the original non-surgical method, it now accounts for approximately 5% of in vitro fertilization recorded births.

Prior to this, thousands of women who were infertile, had adoption as the only path to parenthood. This set the stage to allow open and candid discussion of embryo donation and transfer.

This breakthrough has given way to the donation of human embryos as a common practice similar to other donations such as blood and major organ donations.

At the time of this announcement the event was captured by major news carriers and fuelled healthy debate and discussion on this practice which impacted the future of reproductive medicine by creating a platform for further advancements in woman's health.

This work established the technical foundation and legal-ethical framework surrounding the clinical use of human oocyte and embryo donation, a mainstream clinical practice, which has evolved over the past 25 years. Building upon this groundbreaking research and since the initial birth announcement in 1984, well over 47,000 live births resulting from donor embryo transfer have been and continue to be recorded by the Centres for Disease Control(CDC) in the United States to infertile women, who otherwise would not have had children by any other existing method.

Embryo Transfer in Animals

Embryo transfer techniques allow top quality female livestock to have a greater influence on the genetic advancement of a herd or flock in much the same way that artificial insemination has allowed greater use of superior sires.

ET also allows the continued use of animals such as competition mares to continue training and showing, while producing foals. The general epidemiological aspects of embryo transfer indicates that the transfer of embryos provides the opportunity to introduce genetic material into populations of livestock while greatly reducing the risk for transmission of infectious diseases. Recent developments in the sexing of embryos before transfer and implanting has great potential in the dairy and other livestock industries.

Embryo transfer is also used in laboratory mice. For example, embryos of genetically modified strains that are difficult to breed or expensive to maintain may be stored frozen, and only thawed and implanted into a pseudopregnant dam when needed.

Success Rates

IVF success rates are the percentage of all IVF procedures which result in a favourable outcome. Depending on the type of calculation used, this outcome may represent the number of confirmed pregnancies, called the pregnancy rate or number of live births, called the live birth rate. Due to advancement in reproductive technology, the IVF success rates are substantially better today than they were just a few years ago. The most current data available in the United States a 2009 summary complied by the Society for Reproductive Medicine (SART) which reports the average national IVF success rates per age group using non-donor eggs.

	<35	*35-37*	*38-40*	*41-42*
Pregnancy Rate	47.6	38.9	30.1	20.5
Live Birth Rate	41.4	31.7	22.3	12.6

The live birth rates using donor eggs are also given by the SART and include all age groups using either fresh or thawed eggs.

	Fresh Donor Egg Embryos	*Thawed Donor Egg Embryos*
Live Birth Rate	55.1	33.8

In 2006, Canadian clinics reported an average pregnancy rate of 35%. A French study estimated that 66% of patients starting IVF treatment finally succeed in having a child (40% during the IVF treatment at the centre and 26% after IVF discontinuation). Achievement of having a child after IVF discontinuation was mainly due to adoption (46%) or spontaneous pregnancy (42%).

Live Birth Rate

The live birth rate is the percentage of all IVF cycles that lead to a live birth. This rate does not include miscarriage or stillbirth and multiple-order births such as twins and triplets are counted as one pregnancy. The live birth rate is calculated as follows:

Number of Live Births : Number of Procedures Performed

In 2006, Canadian clinics reported a live birth rate of 27%. Birth rates in younger patients were slightly higher, with a success rate of 35.3% for those 21 and younger, the youngest group evaluated. Success rates for older patients were also lower and decrease with age, with 37-year-olds at 27.4% and no live births for those older than 48, the oldest group evaluated.

Some clinics exceeded these rates, but it is impossible to determine if that is due to superior technique or patient selection, because it is possible to artificially increase success rates by refusing to accept the most difficult patients or by steering them into oocyte donation cycles (which are compiled separately). Further, pregnancy rates can be increased by the placement of several embryos at the risk of increasing the chance for multiples.

Because not each IVF cycle that is started will lead to oocyte retrieval or embryo transfer, reports of live birth rates need to specify the denominator, namely IVF cycles started, IVF retrievals, or embryo transfers. The Society for Assisted Reproductive Technology (SART) summarised 2008-9 success rates for US clinics for fresh embryo cycles that did not involve donor eggs and gave live birth rates by the age of the prospective mother, with a peak at 41.3% per cycle started and 47.3% per embryo transfer for patients under 35 years of age.

IVF attempts in multiple cycles result in increased cumulative live birth rates. Depending on the demographic group, one study reported 45% to 53% for three attempts, and 51% to 71% to 80% for six attempts.

Pregnancy Rate

Pregnancy rate may be defined in various ways. In the United States, the pregnancy rate used by the Society for Assisted Reproductive Technology and the Centres for Disease Control (and appearing in the table in the Success Rates section above) are based on fetal heart motion observed in ultrasound examinations.

However, some clinics may advertise pregnancy rates based on pregnancy tests which are less definitive, such as urine or blood tests for hCG. This is the rate most often published by fertility clinics because often at the time of publishing not all IVF patients have given birth. The pregnancy rate is calculated as follows:

Number pregnancies / Number of procedures performed (cycle start, egg retrieval or embryo transfer).

Success or Failure Factors

The main potential factors that influence pregnancy (and live birth) rates in IVF have been suggested to be maternal age, duration of infertility or subfertility, bFSH and number of oocytes, all reflecting ovarian function. Optimal woman's age is 23–39 years at time of treatment.

Stress

In a 2005 Swedish study, 166 women were monitored starting one month before their IVF cycles, and the results showed no significant correlation between psychological stress and IVF outcome. The study concluded with the recommendation to clinics that it might be possible to reduce the stress experienced by IVF patients during the treatment procedure by informing them of those findings. While psychological stress experienced during a cycle might not influence an IVF outcome, it is possible that the experience of IVF can result in stress that leads to depression. The financial consequences alone of IVF can influence anxiety and become overwhelming. However, for many couples, the alternative is infertility, and the experience of infertility itself can also cause extreme stress and depression.

Acupuncture

An increasing number of fertility specialists and centres offer acupuncture as a part of their IVF protocol. Limited but supportive evidence from clinical trials and case series suggests that acupuncture may improve the success rate of IVF and the quality of life of patients undergoing IVF and that it is a safe adjunct therapy. A systematic review and meta-analysis published in the British Medical Journal found that complementing the embryo transfer process with acupuncture was associated with significant and clinically relevant improvements in clinical pregnancy (where the expected number of patients needed to be treated to produce 1 additional pregnancy was 10), ongoing pregnancy (NNT 9), and live birth (NNT 9).

Acupuncture Mechanisms

Four mechanisms by which it has been suggested that acupuncture may improve IVF outcomes are

- Neuroendocrinological modulations
- Increased blood flow to uterus and ovaries
- Modulation in cytokines
- Reduction of stress, anxiety, and depression.

Electro-acupuncture in Oocyte Retrieval for IVF

Electro-acupuncture has been found to be a good alternative to conventional medical analgesia, it results in shorter hospitalisation times and lower costs.

Other Factors

Other determinants of outcome of IVF include:

- Tobacco smoking reduces the chances of IVF producing a live birth by 34% and increases the risk of an IVF pregnancy miscarrying by 30%.
- A body mass index (BMI) over 27 causes a 33% decrease in likelihood to have a live birth after the first cycle of IVF, compared to those with a BMI between 20 and 27. Also, pregnant women who are obese have higher rates of congenital abnormality, miscarriage, gestational diabetes, hypertension, thromboembolism and problems during delivery. Ideal body mass index is 19–30.
- Salpingectomy before IVF treatment increases chances for women with hydrosalpinges
- Success with previous pregnancy and/or live birth increases chances
- Low alcohol/caffeine intake increases success rate
- Level of DNA fragmentation as measured e.g. by Comet assay, advanced maternal age and semen quality.
- The number of embryos transferred in the treatment cycle.
- Other factors of semen quality for the sperm provider.
- Women with ovary-specific FMR1 genotypes including *het-norm/low* have significantly decreased pregnancy chances in IVF.
- Some studies also suggest the autoimmune disease may also play a role in decreasing IVF success rates by interfering with proper implantation of the embryo after transfer.

Aspirin, however, was shown by a meta-analysis to not improve pregnancy rates after IVF.

Risks

Possible risks may occur throughout the procedure, but the risk you may encounter depend on the specific step of the procedure. During ovarian stimulation, hyperstimulation syndrome may occur. This results in swollen, painful ovaries and some form of it (mild, moderate or severe) occurs in 30% of patients. Mild cases can be treated with over the counter medications and cases can be resolved in the absence of pregnancy. In moderate cases, ovaries swell and fluid accumulated in the abdominal cavities and may have symptoms of heartburn, gas, nausea or loss of appetite. In severe cases patients have sudden excess abdominal pain, nausea, vomiting and will result in hospitalisation.

During egg retrieval, there's a small chance of bleeding, infection, and damage to surrounding structures like bowel and bladder (transvaginal ultrasound aspiration) as well as difficulty breathing, chest infection, allergic reactions to meds, or nerve damage (laproscopy). During embryo transfer, if more than one embryo is transferred there's always a risk of multiple pregnancy, infertile couples may see this is good news but there may be risk to the embryos and to the mother such as premature delivery. Ectopic pregnancy may also occur if a fertilized egg develops outside the uterus, usually in the fallopian tubes and requires immediate destruction of the fetus.

Complications

Multiple Births

The major complication of IVF is the risk of multiple births. This is directly related to the practice of transferring multiple embryos at embryo transfer. Multiple births are related to increased risk of pregnancy loss, obstetrical complications, prematurity, and neonatal morbidity with the potential for long term damage. Strict limits on the number of embryos that may be transferred have been enacted in some countries (e.g. Britain, Belgium) to reduce the risk of high-order multiples (triplets or more), but are not universally followed or accepted. Spontaneous splitting of embryos in the womb after transfer can occur, but this is rare and would lead to identical twins.

A double blind, randomised study followed IVF pregnancies that resulted in 73 infants (33 boys and 40 girls) and reported that 8.7% of singleton infants and 54.2% of twins had a birth weight of $<$ 2,500 grams (5.5 lb). However, despite the risks, some recent evidence suggests that twins conceived by IVF actually have a 40% decreased outcome risk for complications than spontaneous twin conceptions. Experts believe that this could be due to high level of prenatal monitoring received from the moment of conception. Recent evidence also suggest that singleton offspring after IVF is at higher risk for lower birth weight for unknown reasons.

Birth Defects

The issue of birth defects has been a controversial topic in IVF. Many studies do not show a significant increase after use of IVF, and some studies suggest higher rates for ICSI, whereas others do not support this finding.

In 2008, an analysis of the data of the National Birth Defects Study in the US found that certain birth defects were significantly more common in infants conceived through IVF, notably septal heart defects, cleft lip with or without cleft palate, esophageal atresia, and anorectal atresia; the mechanism of causality is unclear.

A 2002 study, reviewing the birth records of all the IVF babies in Western Australia "has found IVF kids are twice as likely to have birth defects as normal births". The study allowed for the effects of multiple births and the mother's age.

The higher incidence of birth defects "up to one year of age" included: heart defects, "chromosomal abnormalities like Down syndrome, spina bifida, gastro-intestinal abnormalities, musculo-skeletal, dislocated hips," and club feet.

The IVF children also were found to have a higher incidence of lower birth weights, pre-term births, and cerebral palsy. Japan's government prohibited the use of *in vitro* fertilization procedures for couples in which both partners are infected with HIV.

Despite the fact that the ethics committees previously allowed the Ogikubo, Tokyo Hospital, located in Tokyo, to use *in vitro* fertilization for couples with HIV, the Ministry of Health, Labour and Welfare of Japan decided to block the practice.

Hideji Hanabusa, the vice president of the Ogikubo Hospital, states that together with his colleagues, he managed to develop a method through which scientists are able to remove HIV from sperm.

Other Risks

Another risk of ovarian stimulation is the development of ovarian hyperstimulation syndrome, particularly if hCG is used to "trigger ovulation".

If the underlying infertility is related to abnormalities in spermatogenesis, it is plausible, but too early to examine that male offspring is at higher risk for sperm abnormalities. Behaviour and socioemotional functioning of children conceived by IVF is normal overall, according to studies on 9–18-year-old IVF children.

A negative pregnancy test after IVF is associated with an increased risk for depression in women, but not with any increased risk of developing anxiety disorders. Pregnancy test results do not seem to be a risk factor for depression or anxiety among men.

Expansions

There are various expansions or additional techniques that can be applied in IVF, which are usually not necessary for the IVF procedure itself, but would be virtually impossible or technically difficult to perform without concomitantly performing methods of IVF.

Embryo Profiling

In addition to the methods already mentioned in the *embryo selection* above, there are more advanced techniques of *embryo profiling* that give more information to aid in the selection of the optimal embryo.

Also, methods are emerging in making comprehensive analyses of up to entire genomes, transcriptomes, proteomes and metabolomes which may be used to score embryos by comparing the patterns with ones that have previously been found among embryos in successful versus unsuccessful pregnancies:

- In transcriptome evaluation, however, gene expression profiling studies of human embryos are limited due to legal and ethical issues. An alternative or complement is gene expression profiling of cumulus cells surrounding the oocyte and early embryo, or on granulosa cells. Profiling of cumulus cells can give valuable information regarding the efficiency of an ovarian hyperstimulation protocol, and may indirectly predict oocyte aneuploidy, embryo development and pregnancy outcomes, without having to perform any invasive procedure directly in the embryo.
- Proteome profiling of embryos can indirectly be evaluated by sampling of proteins found in the vicinity of embryos, thereby providing a non-invasive method of embryo profiling. Examples of protein markers evaluated in such profiling include CXCL13 and granulocyte-macrophage colony-stimulating factor, where lower protein amounts are associated with higher implantation rates.

Another level of opportunity can be achieved by having the evaluation of the embryo profile tailored to the maternal status in regard to, for example health or immune status, potentially further detailed by similar profiling of the maternal genome, transcriptome, proteome and metabolome. Two examples of proteins that may be included in maternal profiling are endometrium-derived stathmin 1 and annexin A2, whose down- and up-regulation, respectively, are associated with higher rates of successful implantation.

The end-point of interest in the evaluation of an embryo profile is not necessarily the success rate of the pregnancy at hand, but may also be the health of the potential person as a child and adult in the future, such as in detection of inheritable diseases by preimplantation genetic diagnosis (PGD) or screening (PGS):

Preimplantation Genetic Screening or Diagnosis (PGS or PGD)

Preimplantation genetic screening (PGS) or preimplantation genetic diagnosis (PGD) has been suggested to be able to be used in IVF to select an embryo that appears to have the greatest chances for successful pregnancy. However, a systematic review and meta-analysis of existing randomised controlled trials came to the result that there is no evidence of a beneficial effect of PGS as measured by live birth rate.

On the contrary, for women of advanced maternal age, PGS significantly lowers the live birth rate. Technical drawbacks, such as the invasiveness of the biopsy, and non-representative samples because of mosaicism are the major underlying factors for inefficacy of PGS.

Still, as an expansion of IVF, patients who can benefit from PGS/PGD include:

- Couples who have a family history of inherited disease
- Couples who want to use gender selection to prevent a gender-linked disease
- Couples who already have a child with an incurable disease and need compatible cells from a second healthy child to cure the first, resulting in a "saviour sibling" that matches the sick child in HLA type.

PGS screens for numeral chromosomal abnormalities while PGD diagnosis the specific molecular defect of the inherited disease. In both PGS and PGD, individual cells from a pre-embryo are analysed during the IVF process.

Before the transfer of a pre-embryo back to a woman's uterus, one or two cells are removed from the pre-embryos (8-cell stage). These cells are then evaluated for normality. Typically within one to two days, following completion of the evaluation, only the normal pre-embryos are transferred back to the woman's uterus. In addition, PGS can reduce the risk of multiple pregnancies because fewer embryos are needed for implantation.

Cryopreservation

Cryopreservation can be performed as oocyte cryopreservation before fertilization, or as embryo cryopreservation after fertilization. The Rand Consulting Group has estimated there to be 400,000 frozen embryos in the

United States. The advantage is that patients who fail to conceive may become pregnant using such embryos without having to go through a full IVF cycle. Or, if pregnancy occurred, they could return later for another pregnancy.

Spare oocytes or embryos resulting from fertility treatments may be used for oocyte donation or embryo donation to another woman or couple, and embryos may be created, frozen and stored specifically for transfer and donation by using donor eggs and sperm. Also, oocyte cryopreservation can be used for women who are likely to lose their ovarian reserve due to undergoing chemotherapy. The outcome from using cryopreserved embryos has uniformly been positive with no increase in birth defects or development abnormalities.

Bibliography

Altered Fates: *Gene Therapy and the Retooling of Human Life*, W. W. Norton and Company, New York, 1995.

Bellairs, A.: *The Life of Reptiles*, Weidenfield and Nicholson, London, 1969.

Curan, Brendan: *A Terrible Beauty is Born: Clones, Genes and the Future of Mankind*. New York: Taylor & Francis, 2003.

Dhawan, Ruchi: *Cloning, Genetics and Stem Cell Technology : Development and Applications*, Axis Pub, Delhi, 2010.

Fisher, R. A.: *The Genetical Theory of Natural Selection*, Clarendon Press, Oxford, 1930.

Friedberg, E.C.: *DNA Repair*, New York, WH Freeman and Company, 1985.

Gall, Joseph G.: *Landmark Papers in Cell Biology,* Cold Spring Harbor Laboratory Press, Plainview, NY, 2001.

Glover, D. M., and Hames, B. D.: *Genes and Embryos*, Oxford University Press, New York, 1989.

Hartl, Daniel L.: *Basic Genetics*, Jones and Bartlett Publishers, Boston, 1991.

Jones, Gareth D., and Mary Byrne: *Stem Cell Research and Cloning: Contemporary Challenges to Our Humanity*. Hindmarsh, SA, Australia: Australasian Theological Forum, 2005.

Kass, Leon R and James Q Wilson: *The Ethics of Human Cloning*, Scientia, Delhi, 2002.

Kumar, Rajesh: *Fundamental of Artificial Neural Network and Fuzzy Logic*, Laxmi Pub, Delhi, 2009.

Lamba, V.K.: *Neuro Fuzzy Systems*, Laxmi Publications, Delhi, 2008.

Lawrence, Jeanette: *Introduction to Neural Networks*, California Scientific Software Press, 1994.

Margulis, L.: *Symbiosis in Cell Evolution,* W.H. Freeman, San Francisco, 1981.

Roetz, Heiner: *Cross-cultural Issues in Bioethics: The Example of Human Cloning.* Amsterdam: Rodopi, 2006.

Ross T.: *Fuzzy Logic with Engineering Applications.* McGraw-Hill, New York, 1995.

Scriver, C.R.: *The Metabolic and Molecular Basis of Inherited Disease,* McGraw Hill, New York, 2001.

Silver, Lee M: *Remaking Eden: How Genetic Engineering and Cloning Will Transform the American Family.* New York: Avon, 1998.

Stoll R.R.: *Set Theory and Logic,* Dover Publications, New York, 1979.

Sugeno, M.: *Industrial Applications of Fuzzy Control,* Elsevier Science Pub. Co., 1985.

Von Altrock, Constantin: *Fuzzy Logic and NeuroFuzzy Applications Explained.* Upper Saddle River, NJ: Prentice Hall, 1995.

Wilmut, Ian and Roger Highfield: *After Dolly: The Uses and Misuses of Human Cloning.* New York: W. W. Norton, 2006.

Wilson, E. O. and Lumden, C.: *Genes, Mind, and Culture: The Evolutionary Process,* Harvard University Press, Cambridge, Mass, 1981.

Woodward, John: *The Ethics of Human Cloning.* San Diego: Greenhaven Press, 2004.

Wright, L.: *Twins: Genes, Environment and the Mystery of Human Identity,* Weidenfeld & Nicolson, London, 1997.

Index

❑❑❑